Textile Printing

Textile Printing

Dr. N.N. Mahapatra

*B.Sc (Hons) B.Sc (Tech) (Bom) M.Sc (Chem), Ph.D (Chem), M.B.A
(IMM,Cal) C.Col FSDC (UK), CText FTI (Manchester), FRSC
(UK) Int Trg (Australia)), Sen Mem, AATCC (USA), FAIC(USA)
FIC, FTA, FICS, FIE, FIIChE, MISTE (India)*

CRC Press
Taylor & Francis Group
Boca Raton London New York

CRC Press is an imprint of the
Taylor & Francis Group, an **informa** business

WOODHEAD PUBLISHING INDIA PVT LTD

New Delhi

First edition published 2024
by CRC Press
4 Park Square, Milton Park, Abingdon, Oxon, OX14 4RN

and by CRC Press
2385 NW Executive Center Drive, Suite 320, Boca Raton FL 33431

British Library Cataloguing-in-Publication Data
A catalogue record for this book is available from the British Library

Print edition not for sale in South Asia (India, Sri Lanka, Nepal, Bangladesh, Pakistan or Bhutan)

ISBN: 9781032630083 (hbk)
ISBN: 9781032630106 (pbk)
ISBN: 9781032630113 (ebk)

DOI: 10.4324/9781032630113

Times New Roman
by Bhumi Graphics, New Delhi

Contents

Preface

During 1981-84, while I was studying at UDCT, now known as ICT, in Matunga, Mumbai, I remember one of the textbooks on Textile Printing written by our Professor and my guide Dr V.A.Shenai. As I recall, it was indeed the only book available on Textile Printing. Prof Shenai used to teach us about the subject in a very lucid manner. Learning from him was exciting and very informative as he was also doing consultancy for many Textile Mills in and around Mumbai, which would fetch us practical information during his lectures. Prof Mangesh Teli and Prof S.R.Shukla used to guide us in the lab to perform practicals on Textile Printing.

I was admiringly impressed by Prof Shenai as a person, writer, and teacher. After writing my book on Textile Dyes And Textile Dyeing, I was intrigued to write a book on Textile Printing. The people I have encountered in my life have left a positive mark on me; Prof Shenai has written more than 20 textile books. My father, Late Prof Dr Gokulananda Mahapatra, has also written more than 90 books on Chemistry, Science Novels and Science Fiction. They have been the inspiration that has motivated me to write books and collate my industrial and practical knowledge to life.

In all my books, I share more than 25 years of experience working on the shop floor in various Textile mills in India and abroad. In this book, I have incorporated new topics like Digital Printing, Yarn Printing, Transfer Printing and more.

It's about time we keep updating and upgrading ourselves, and so should our printing knowledge because we are entering an era with Water Less Dyeing, Air Dyeing and Salt-Free Reactive Dyeing.

I am thankful to my daughters Nittisha (Copywriter and communication strategist) and Sanchitta (Studying Master's in Design from Arizona State University, USA), for their timely assistance in compiling this book's manuscript.

I want to dedicate this book to my dear wife, Seemani, for her endless support in completing this book.

Nanda Nandan Mahapatra
Place- Mumbai
Date – 8 th July, 2022

1
Introduction

Textile printing is the process of applying colour to fabric in definite patterns or designs. In properly printed fabrics the colour is bonded with the fibre, so as to resist washing and friction. Textile printing is related to dyeing but in dyeing properly the whole fabric is uniformly covered with one colour, whereas in printing one or more colours are applied to it in certain parts only, and in sharply defined patterns.

In printing, wooden blocks, stencils, engraved plates, rollers, or silkscreens can be used to place colours on the fabric. Colourants used in printing contain dyes thickened to prevent the colour from spreading by capillary attraction beyond the limits of the pattern or design.

Woodblock printing is a technique for printing text, images or patterns used widely throughout East Asia and probably originating in China in antiquity as a method of printing on textiles and later paper. As a method of printing on cloth, the earliest surviving examples from China date to before 220.

Textile printing was known in Europe, via the Islamic world, from about the 12th century, and widely used. However, the European dyes tended to liquify, which restricted the use of printed patterns. Fairly large and ambitious designs were printed for decorative purposes such as wall-hangings and lectern-cloths, where this was less of a problem as they did not need washing. When paper became common, the technology was rapidly used on that for woodcut prints.[Superior cloth was also imported from Islamic countries, but this was much more expensive.

The Incas of Peru, Chile and the Aztecs of Mexico also practiced textile printing previous to the Spanish Invasion in 1519; but owing to the lack of records before that date, it is impossible to say whether they discovered the art for themselves, or, in some way, learned its principles from the Asiatics.

During the later half of the 17th century the French brought directly by sea, from their colonies on the east coast of India, samples of Indian blue and white resist prints, and along with them, particulars of the processes by which they had been produced, which produced washable fabrics.

As early as the 1630s, the East India Company was bringing in printed and plain cotton for the English market. By the 1660s British printers and dyers were making their own printed cotton to sell at home, printing single colors on plain backgrounds; less colourful than the imported prints, but more to the taste of the British. Designs were also sent to India for their craftspeople to copy for export back to England. There were many dyehouses in England in the latter half of the 17th century, Lancaster being one area and on the River Lea near London another. Plain cloth was put through a prolonged bleaching process which prepared the material to receive and hold applied color; this process vastly improved the color durability of English calicoes and required a great deal of water from nearby rivers. One dyehouse was started by John Meakins, a London Quaker who lived in Cripplegate. When he died, he passed his dyehouse to his son-in-law Benjamin Ollive, Citizen and Dyer, who moved the dye-works to Bromley Hall where it remained in the family until 1823, known as Benjamin Ollive and Company, Ollive & Talwin, Joseph Talwin & Company and later Talwin & Foster. Samples of their fabrics and designs can be found in the Victoria and Albert Museum in London and the Smithsonian Copper-Hewett in New York.

On the continent of Europe the commercial importance of calico printing seems to have been almost immediately recognized, and in consequence it spread and developed there much more rapidly than in England, where it was neglected for nearly ninety years after its introduction. During the last two decades of the 17th century and the earlier ones of the 18th new dye works were started in France, Germany, Switzerland and Austria. It was only in 1738 that calico printing was first, practiced in Scotland, and not until twenty-six years later that Messrs Clayton of Bamber Bridge, near Preston, established in 1764 the first print-works in Lancashire, and thus laid the foundation of the industry.

From an artistic point of view most of the pioneer work in calico printing was done by the French. From the early days of the industry down to the latter half of the 20th century, the productions of the French printers in Jouy, Beauvais, Rouen, and in Alsace-Lorraine, were looked upon as representing all that was best in artistic calico printing.

1.1   Methods

Traditional textile printing techniques may be broadly categorized into four styles:

- Direct printing, in which colorants containing dyes, thickeners, and the mordants or substances necessary for fixing the color on the cloth are printed in the desired pattern.

- The printing of a mordant in the desired pattern prior to dyeing cloth; the color adheres only where the mordant was printed.
- Resist dyeing, in which a wax or other substance is printed onto fabric which is subsequently dyed. The waxed areas do not accept the dye, leaving uncolored patterns against a colored ground.
- Discharge printing, in which a bleaching agent is printed onto previously dyed fabrics to remove some or all of the colour.

Resist and discharge techniques were particularly fashionable in the 19th century, as were combination techniques in which indigo resist was used to create blue backgrounds prior to block-printing of other colors.[2] Modern industrial printing mainly uses direct printing techniques.

The printing process does involve several stages in order to prepare the fabric and printing paste, and to fix the impression permanently on the fabric:

- pre-treatment of fabric,
- preparation of colors,
- preparation of printing paste,
- impression of paste on fabric using printing methods,
- drying of fabric,
- fixing the printing with steam or hot air (for pigments),
- after process treatments.

1.2 Preparation of cloth for printing

The fabric is prepared by washing and bleaching. For a coloured ground it is then dyed. The fabric has always to be brushed, to free it from loose nap, flocks and dust that it picks up whilst stored. Frequently, too, it has to be sheared by being passed over rapidly revolving knives arranged spirally round an axle, which rapidly and effectually cuts off all filaments and knots, leaving the fabric perfectly smooth and clean and in a condition fit to receive impressions of the most delicate engraving. Some fabrics require very careful stretching and straightening on a stenter before they are wound around hollow wooden or iron centers into rolls of convenient size for mounting on the printing machines.

1.3 Preparation of colours

The art of making colours for textile printing demands both chemical knowledge and extensive technical experience, for their ingredients must not only be in proper proportion to each other, but also specially chosen and compounded for the particular style of work in hand. A colour must comply to conditions such as shade, quality and fastness; where more colours are associated in the same design each must be capable of withstanding the various operations necessary for the development and fixation of the others. All printing pastes whether containing colouring matter or not are known technically as colours.

Colours vary considerably in composition. Most of them contain all the elements necessary for direct production and fixation. Some, however, contain the colouring matter alone and require various after-treatments; and others again are simply thickened mordants. A mordant is a metallic salt or other substance that combines with the dye to form an insoluble colour, either directly by steaming, or indirectly by dyeing. All printing colours require thickening to enable them to be transferred from colour-box to cloth without running or spreading beyond the limits of the pattern.

1.4   Thickening agents

The printing thickeners used depend on the printing technique, the fabric and the particular dyestuff . Typical thickening agents are starch derivatives, flour, gum arabic, guar gum derivatives, tamarind, sodium alginate, sodium polyacrylate, gum Senegal and gum tragacanth, British gum or dextrine and albumen.

Hot-water-soluble thickening agents such as native starch are made into pastes by boiling in double or jacketed pans. Most thickening agents used today are cold-soluble and require only extensive stirring.

1.5   Starch paste

Starch paste is made from wheat starch, cold water, and olive oil, then thickened by boiling. Non-modified starch is applicable to all but strongly alkaline or strongly acid colours. With the former it thickens up to a stiff unworkable jelly. In the case of the latter, while mineral acids or acid salts convert it into dextrine, thus diminishing its viscosity or thickening power, organic acids do not have that effect. Today, modified carboxymethylated cold soluble starches are mainly used. These have a stable viscosity and are easy to rinse out of the fabric and give reproducible «short» paste rheology.

Flour paste is made in a similar way to starch paste; it is sometimes used to thicken aluminum and iron mordants. Starch paste resists of rice flour have been used for several centuries in Japan.

1.6 Gums

Gum arabic and gum Senegal are both traditional thickenings, but expense prevents them from being used for any but pale, delicate tints. They are especially useful thickenings for the light ground colours of soft muslins and sateens on account of the property they possess of dissolving completely out of the fibres of the cloth in the post-printing washing process, and they have a long flowing, viscous rheology, giving sharp print and good penetration in the cloth. Today guar gum and tamarind derivates offer a cheaper alternative.

British gum or dextrin is prepared by heating starch. It varies considerably in composition, sometimes being only slightly roasted and consequently only partly converted into dextrine, and at other times being highly torrefied, and almost completely soluble in cold water and very dark in colour. Its thickening power decreases and its gummy nature increases as the temperature at which it is roasted is raised. It is useful for strongly acid colours, and with the exception of gum Senegal, it is the best choice for strongly alkaline colours and discharges. Like the natural gums, it does not penetrate as well into the fibre of the cloth pr as deeply as pure starch or flour and is unsuitable for very dark, strong colours.

Gum tragacanth, or Dragon, which may be mixed in any proportion with starch or flour, is equally useful for pigment colours and mordant colours. When added to a starch paste it increases its penetrative power and adds to its softness without diminishing its thickness, making it easier to wash out of the fabric. It produces much more even colours than does starch paste alone. Used by itself it is suitable for printing all kinds of dark grounds on goods that are required to retain their soft clothy feel.

Starch always leaves the printed cloth somewhat harsh in feeling (unless modified carboxymethylated starches are used), but very dark colours can be obtained. Gum Senegal, gum arabic or modified guar gum thickening yield clearer and more even tints than does starch, suitable for lighter colours but less suited for very dark colours. (The gums apparently prevent the colours from combining fully with the fibers.) A printing stock solution is mostly a combination of modified starch and gum stock solutions.

1.7 Albumen

Albumen is both a thickening and a fixing agent for insoluble pigments. Chrome yellow, the ochres, vermilion and ultramarine are such pigments. Albumen is always dissolved in the cold, a process that takes several days when large quantities are required. Egg albumen is expensive and only used for the lightest shades. Blood albumen solution is used in cases when very dark colours are required to be absolutely fast to washing. After printing, albumen thickened colours are exposed to hot steam, which coagulates the albumen and effectually fixes the colours.

1.8   Printing paste preparation

Combinations of cold water-soluble carboxymethylated starch, guar gum and tamarind derivatives are most commonly used today in disperse screen printing on polyester. Alginates are used for cotton printing with reactive dyes, sodium polyacrylates for pigment printing, and in the case of vat dyes on cotton only carboxymethylated starch is used.

Formerly, colors were always prepared for printing by boiling the thickening agent, the colouring matter and solvents, together, then cooling and adding various fixing agents. At the present time, however, concentrated solutions of the colouring matters and other adjuncts are often simply added to the cold thickenings, of which large quantities are kept in stock.

Colours are reduced in shade by simply adding more stock (printing) paste. For example, a dark blue containing 4 oz. of methylene blue per gallon may readily be made into a pale shade by adding to it thirty times its bulk of starch paste or gum, as the case may be. The procedure is similar for other colours.

Before printing it is essential to strain or sieve all colours in order to free them from lumps, fine sand, and other impurities, which would inevitably damage the highly polished surface of the engraved rollers and result in bad printing. Every scratch on the surface of a roller prints a fine line on the cloth, and too much care, therefore, cannot be taken to remove, as far as possible, all grit and other hard particles from every colour.

Straining is usually done by squeezing the colour through filter cloths like artisanal fine cotton, silk or industrial woven nylon. Fine sieves can also be employed for colours that are used hot or are very strongly alkaline or acid.

1.9   Methods of printing

There are eight distinct methods presently used to impress coloured patterns on cloth:
- Hand block printing
- Perrotine printing
- Engraved copperplate printing

- Roller, cylinder, or machine printing
- Stencil printing
- Screen printing
- Digital textile printing
- Flexo textile printing
- Discharge Printing

Hand block process is the earliest, simplest and slowest of all printing methods. A design is drawn on, or transferred to, prepared wooden blocks. A separate block is required for each distinct colour in the design. A blockcutter carves out the wood around the heavier masses first, leaving the finer and more delicate work until the last so as to avoid any risk of injuring it when the coarser parts are cut. When finished, the block has the appearance of a flat relief carving, with the design standing out. Fine details, difficult to cut in wood, are built up in strips of brass or copper, which is bent to shape and driven edgewise into the flat surface of the block. This method is known as coppering.

The printer applies colour to the block and presses it firmly and steadily on the cloth, striking it smartly on the back with a wooden mallet. The second impression is made in the same way, the printer taking care to see that it registers exactly with the first. Pins at each corner of the block join up exactly, so that the pattern can continue without a break. Each succeeding impression is made in precisely the same manner until the length of cloth is fully printed. The cloth is then wound over drying rollers. If the pattern contains several colours the cloth is first printed throughout with one color, dried, and then printed with the next.

Block printing by hand is a slow process. It is, however, capable of yielding highly artistic results, some of which are unobtainable by any other method. William Morris used this technique in some of his fabrics.

The perrotine is a block-printing machine invented by Perrot of Rouen in 1834 and is now only of historical interest.

This process was patented by Thomas Bell in 1785, fifteen years after his use of an engraved plate to print textiles. Bell's patent was for a machine to print six colours at once, but, probably owing to its incomplete development, it was not immediately successful. One colour could be printed with satisfactorily; the difficulty was to keep the six rollers in register with each other. This defect was overcome by Adam Parkinson of Manchester in 1785. That year, Bells machine with Parkinson›s improvement was successfully employed by Messrs Livesey, Hargreaves and Company of Bamber Bridge, Preston, for the printing of calico in from two to six colours at a single operation.

Roller printing was highly productive, 10,000 to 12,000 yards being commonly printed in one day of ten hours by a single-colour machine. It is capable of reproducing every style of design, ranging from the fine delicate lines of copperplate engraving to the small repeats and limited colours of the perrotine to the broadest effects of block printing with repeats from 1 in to 80 inches. It is precise, so each portion of an elaborate multicolour pattern can be fitted into its proper place without faulty joints at the points of repetition.

1.10   Stencil printing

The art of stenciling on textile fabrics has been practiced from time immemorial by the Japanese, and found increasing employment in Europe for certain classes of decorative work on woven goods during the late 19th century.

A pattern is cut from a sheet of stout paper or thin metal with a sharp-pointed knife, the uncut portions representing the part that will be left uncoloured. The sheet is laid on the fabric and colour is brushed through its interstices.

The peculiarity of stenciled patterns is that they have to be held together by ties. For instance, a complete circle cannot be cut without its centre dropping out, so its outline has to be interrupted at convenient points by ties or uncut portions. This limitation influences the design.

For single-colour work a stenciling machine was patented in 1894 by S. H. Sharp. It consists of an endless stencil plate of thin sheet steel that passes continuously over a revolving cast iron cylinder. The cloth to be ornamented passes between the two and the colour is forced onto it through the holes in the stencil by mechanical measure.

Screen printing is by far the most common technology today. Two types exist: rotary screen printing and flat (bed) screen printing. A blade (squeegee) squeezes the printing paste through openings in the screen onto the fabric.

Digital textile printing is often referred to as direct-to-garment printing, DTG printing, or digital garment printing. It is a process of printing on textiles and garments using specialized or modified inkjet technology. Inkjet printing on fabric is also possible with an inkjet printer by using fabric sheets with a removable paper backing. Today, major inkjet technology manufacturers can offer specialized products designed for direct printing on textiles, not only for sampling but also for bulk production. Since the early 1990s, inkjet technology and specially developed water-based ink (known as dye-sublimation or disperse direct ink) have made it possible to print directly onto polyester fabric. This is mainly related to visual communication in retail and brand promotion (flags, banners and other point of sales applications). Printing onto nylon and silk can be done by using an acid ink. Reactive ink is used for cellulose based fibers such as cotton and linen. Inkjet technology in digital textile printing allows for single pieces, mid-run production and even long-run alternatives to screen printed fabric.

Flexo textile printing on textile fabric was success in China in the last 4 years. Central Impression Flexo, Rubber Sleeves as the printing plate in round engraved by laser(Direct Laser Engraving), Anilox in Sleeve technologies are applicated in the area. Not only the solid, but also 6 to 8 colors in fine register, higher resolution ratio & higher productivity which are the outstanding advantages extraordinary different from the traditional screen textile printing. Aerospace Huayang, Hell system, SPGprints, & Felix Böttcher contributed their technologies & efforts.

1.11   Other methods of printing

Although most work is executed throughout by one or another of the seven distinct processes mentioned above, combinations are frequently employed. Sometimes a pattern is printed partly by machine and partly by block, and sometimes a cylindrical block is used along with engraved copper-rollers in an ordinary printing machine. The block in this latter case is in all respects, except for shape, identical with a flat wood or coppered block, but, instead of being dipped in colour, it receives its supply from an endless blanket, one part of which works in contact with colour-furnishing rollers and the other part with the cylindrical block. This block is known as a surface or peg roller. Many attempts have been made to print multicolour patterns with surface rollers alone, but hitherto with little success, owing to their irregularity in action and to the difficulty of preventing them from warping. These defects are not present in the printing of linoleum in which opaque oil colours are used, colours that neither sink into the body of the hard linoleum nor tend to warp the roller.

Lithographic printing has been applied to textile fabrics with qualified success. Its irregularity and the difficulty of registering repeats have restricted its use to the production of decorative panels, equal or smaller in size to the plate or stone.

Pad printing has been recently introduced to textile printing for the specific purpose of printing garment tags and care labels.

1.12   Calico printing

Goods intended for calico printing are well-bleached; otherwise stains and other serious defects are certain to arise during subsequent operations.

The chemical preparations used for special styles will be mentioned in their proper places; but a general prepare, employed for most colours that are developed and fixed by steaming only, consists in passing the bleached calico through a weak solution of sulphated or turkey red oil containing 2.5 to 5 percent fatty acid. Some colours are printed on pure bleached cloth, but all patterns containing alizarine red, rose and salmon shades are considerably brightened by the presence of oil, and indeed very few, if any, colours are detrimentally affected by it.

The cloth is always brushed to free it from loose nap, flocks and dust that it picks up whilst stored. Frequently, too, it has to be sheared by being passed over rapidly revolving knives arranged spirally round an axle, which rapidly and effectually cuts off all filaments and knots, leaving the cloth perfectly smooth and clean. It is then stentered, wound onto a beam, and mounting on the printing machines.

The colours and methods employed are the same as for wool, except that in the case of silk no preparation of the material is required before printing, and ordinary dry steaming is preferable to damp steaming.

Both acid and basic dyes play an important role in silk printing, which for the most part is confined to the production of articles for fashion goods, handkerchiefs, and scarves, all articles for which bright colours are in demand. Alizarine and other mordant colours are mainly used for any goods that have to resist repeated washings or prolonged exposure to light. In this case the silk frequently must be prepared in alizarine oil, after which it is treated in all respects like cotton, namely steamed, washed and soaped, the colours used being the same.

Silk is especially adapted to discharge and reserve effects. Most of the acid dyes can be discharged in the same way as when they are dyed on wool. Reserved effects are produced by printing mechanical resists, such as waxes and fats, on the cloth and then dyeing it in cold dye-liquor. The great affinity of the silk fibre for basic and acid dyestuffs enables it to extract colouring matter from cold solutions and permanently combine with it to form an insoluble lake. After dyeing, the reserve prints are washed, first in cold water to remove any colour not fixed onto the fibre, and then in hot water or benzene to dissolve out the resisting bodies.

After steaming, silk goods are normally only washed in hot water, but those printed entirely in mordant dyes will stand soaping, and indeed require it to brighten the colours and soften the material.

Some silk dyes do not require heat setting or steaming. They strike instantly, allowing the designer to dye color upon color. These dyes are intended mostly for silk scarf dyeing. They also dye bamboo, rayon, linen, and some other natural fabrics like hemp and wool to a lesser extent, but do not set on cotton.

Woodblock printing is a technique for printing text, images or patterns used widely throughout East Asia and probably originating in China in antiquity as a method of printing on textiles and later paper. As a method of printing on cloth, the earliest surviving examples from China date to before 220.

Textile printing was known in Europe, via the Islamic world, from about the 12th century, and widely used. However, the European dyes tended to liquify, which restricted the use of printed patterns. Fairly large and ambitious designs were printed for decorative purposes such as wall-hangings and lectern-cloths, where this was less of a problem as they did not need washing. When paper became common, the technology was rapidly used on that for woodcut prints.[1] Superior cloth was also imported from Islamic countries, but this was much more expensive.

The Incas of Peru, Chile and the Aztecs of Mexico also practiced textile printing previous to the Spanish Invasion in 1519; but owing to the lack of records before that date, it is impossible to say whether they discovered the art for themselves, or, in some way, learned its principles from the Asiatics.

During the later half of the 17th century the French brought directly by sea, from their colonies on the east coast of India, samples of Indian blue and white resist prints, and along with them, particulars of the processes by which they had been produced, which produced washable fabrics.

As early as the 1630s, the East India Company was bringing in printed and plain cotton for the English market. By the 1660s British printers and dyers were making their own printed cotton to sell at home, printing single colors on plain backgrounds; less colourful than the imported prints, but more to the taste of the British. Designs were also sent to India for their craftspeople to copy for export back to England. There were many dyehouses in England in the latter half of the 17th century, Lancaster being one area and on the River Lea near London another. Plain cloth was put through a prolonged bleaching process which prepared the material to receive and hold applied color; this process vastly improved the color durability of English calicoes and required a great deal of water from nearby rivers. One dyehouse was started by John Meakins, a London Quaker who lived in Cripplegate. When he died, he passed his dyehouse to his son-in-law Benjamin Ollive, Citizen and Dyer, who moved the dye-works to Bromley Hall where it remained in the family until 1823, known as Benjamin Ollive and Company, Ollive & Talwin, Joseph Talwin & Company and later Talwin & Foster. Samples of their fabrics and designs can be found in the Victoria and Albert Museum in London and the Smithsonian Copper-Hewett in New York.

On the continent of Europe the commercial importance of calico printing seems to have been almost immediately recognized, and in consequence it spread and developed there much more rapidly than in England, where it was neglected for nearly ninety years after its introduction. During the last two decades of the 17th century and the earlier ones of the 18th new dye works were started in France, Germany, Switzerland and Austria. It was only in 1738 that calico printing was first, practiced in Scotland, and not until twenty-six years later that Messrs Clayton of Bamber Bridge, near Preston, established in 1764 the first print-works in Lancashire, and thus laid the foundation of the industry.

From an artistic point of view most of the pioneer work in calico printing was done by the French. From the early days of the industry down to the latter half of the 20th century, the productions of the French printers in Jouy, Beauvais, Rouen, and in Alsace-Lorraine, were looked upon as representing all that was best in artistic calico printing.

1.13   Printing process

Applying coloured patterns and designs to decorate a finished fabric is called 'Printing'. In a proper printed fabric, the colour is affixed to the fiber, so that it may not be affected by washing and friction. Whether a fabric is dyed or printed can be known by examining the outline of the design. On a printed fabric, the outline of a design is sharply defined on the outer side. The design generally do not penetrate to the back of the cloth. However, the design may show up on the reverse side of transparently thin fabrics. These fabrics may be confused with the woven designs where yarn dyed warp and filling are used. If the design is printed on such a fabric, the yarns will show some areas on which colour is not equally distributed.

The Dyes used for printing mostly include vat, reactive, naphthol and disperse colours which have good fastness properties. The pigments, which are not truly dyes, are also used extensively for printing. These colours are fixed to the fiber through resins that are very resistant to laundering or drycleaning. Pigments are among the fastest known colours and are effective for light to medium shades. If used for applying dark colours, they may crock or rub off. Improved resins, better pigments or more effective anticrock agents must be used to solve this problem. Cheap prints are made from basic colours mixed with tartar emetic and tannic acid but they are not acceptable in todays market.

For cotton printing vat and reactive dyes are generally used. Silk is usually printed with acid colours. Wool is printed with acid or chrome dyes but before printing it is treated with chlorine to make it more receptive to colours. Manmade fibers are generally printed with disperse and cationic dyes.

1.14   Methods of printing

Three different approaches or techniques are prevalent for printing colour on a fabric: Direct, Discharge and Resist

1.14.1   Direct printing

It is the most common approach to apply a colour pattern on fabric. It can be done on white or a coloured fabric. If done on coloured fabric, it is known as overprinting. The desired pattern is produced by imprinting dye on the fabric in a paste form. To prepare the print paste, a thickening agent is added to a limited amount of water and dye is dissolved in it. Earlier corn starch was preferred as a thickening agent for cotton printing. Nowadays gums or alginates derived from seaweed are preferred because they are easier to wash out, do not themselves absorb any colour and allow better penetration of colour. Most pigment printing is done without thickeners as the mixing up of resins, solvents and water itself produces thickening.

1.14.2   Discharge printing

In this approach, the fabric is dyed in piece and then it is printed with a chemical that destroys the colour in the designed areas. Sometimes, the base colour is removed and another colour is printed in its place. The printed fabric is steamed and then thoroughly washed. This approach is on decline these days.

1.14.3   Resist printing

In this technique, a resist paste is imprinted on the fabric and then it is dyed. The dye affects only those parts that are not covered by the resist paste. After dyeing, the resist paste is removed leaving a pattern on a dark background.

There are various methods of printing in which one of the above three techniques is used - Block Printing, Roller Printing, Duplex Printing, Stencil

Printing, Screen Printing, Transfer Printing, Blotch Printing, Jet Spray Printing, Electrostatic Printing, Photo Printing, Differential Printing, Warp Printing, Batik Dyeing, Tie Dyeing, Airbrush (Spray) Painting and Digital printing

1.14.3.1 Block printing

The designs are carved on a wooden or metal block and the paste dyestuff is applied to the design on the face of the block. The block is pressed down firmly by hand on the surface of the fabric.

1.14.3.2 Roller printing

In this machine counterpart of block printing, engraved copper cylinders or rollers are used in place of handcarved blocks. With each revolution of the roller, a repeat of the design is printed. The printed cloth is passed into a drying and then a steam chamber where the moisture and heat sets the dye.

1.14.3.3 Duplex printing

Printing is done on both sides of the fabric either through roller printing machine in two operations or a duplex printing machine in a single operation.

1.14.3.4 Screen printing

It is done either with flat or cylindrical screens made of silk threads, nylon, polyester, vinyon or metal. The printing paste or dye is poured on the screen and forced through its unblocked areas onto the fabric. Based on the type of the screen used, it is known as 'Flat Screen Printing' or 'Rotary Screen Printing'.

1.14.3.5 Stencil printing

The design is first cut in cardboard, wood or metal. The stencils may have fine delicate designs or large spaces through which colour is applied on the fabric. Its use is limited due to high costs involved.

1.14.3.6 Transfer printing

The design on a paper is transferred to a fabric by vaporization. There are two main processes for this- Dry Heat Transfer Printing and Wet Heat Transfer Printing. In Conventional Heat Transfer Printing, an electrically heated cylinder is used that presses a fabric against a printed paper placed on a heat resistant blanket. In Infrared Heat Vacuum Transfer Printing, the transfer paper and fabric are passed between infrared heaters and a perforated cylinder which are protected from excessive heat by a shield. The Wet Heat Transfer

Printing uses heat in a wet atmosphere for vaporizing the dye pattern from paper to fabric.

1.14.3.7 Blotch printing

It is a direct printing technique where the background colour and the design are both printed onto a white fabric usually in a one operation. Any of the methods like block, roller or screen may be used.

1.14.3.8 Airbrush (Spray) painting

Designs may be hand painted on fabric or the dye may be applied with a mechanized airbrush which blows or sprays colour on the fabric

1.14.3.9 Electrostatic printing

A dye- resin mixture is spread on a screen bearing the design and the fabric is passed into an electrostatic field under the screen. The dye- resin mixture is pulled by the electrostatic field through the pattern area onto the fabric.

1.14.3.10 Photo printing

The fabric is coated with a chemical that is sensitive to light and then any photograph may be printed on it.

1.14.3.11 Differential printing

It is a technique of printing tufted material made of yarns having different dyeing properties such as carpets. Upto a ten colour effect is possible by careful selection of yarns, dyestuffs and pattern

1.14.3.12 Warp printing

It is roller printing applied to warp yarns before they are woven into fabric.

1.14.3.13 Jet spray printing

Designs are imparted to fabrics by spraying colours in a controlled manner through nozzles.

1.14.3.14 Digital printing

In this form of printing micro-sized droplets of dye are placed onto the fabric through an inkjet printhead. The print system software interprets the data supplied by a cademic_Textiledigital image file. The digital image file has the data to control the droplet output so that the image quality and color control may be achieved. This is the latest development in textile printing and is expanding very fast. Digital Textile Printing

There are five main methods of printing a fabric, these being the block, roller, screen, heat transfer and ink-jet methods. The heat transfer method differs from the others in that it involves the transfer of color from the design printed on paper through the vapour phase into the fibres of the fabric. With the other methods the dye or pigment is applied to the fabric surface through a print paste medium. The ink jet printing process however is a comparatively recent innovation and is referred to as a ‹non-impact› method, because the print paste is fired on to the textile from a jet which is not actually in contact with the fabric.

1.15   Different types of printing method

1.15.1   Block printing

The blocks are usually made of wood and the design is hand carved, so that it stands out in relief against the background surface. The print paste is applied to the design surface on the block and the block then pressed against the fabric. The process is repeated with different designs and colours until the pattern is complete.

Block printing is a slow, laborious process and is not suitable for high volume commercial use. It is a method still practised in the oriental countries where markets exist for the types of printed fabrics produced.

1.15.2   Roller printing

Roller printing has traditionally been preferred for long production runs because of the very high speeds possible. It is also a versatile technique since up to a dozen different colours can be printed simultaneously. The basic roller printing equipment consists of a number of copper faced rollers in which the design is etched. There is a separate printing roller for each colour being printed. Each of the rollers rotates over the fabric under pressure against an iron pressure roller. A blanket and backing cloth rotate over the pressure roller under the fabric and provide a flexible support for the fabric being printed. A colour doctor blade removes paste or fibres adhering to the roller after contact with the fabric. After the impression stage the fabric passes to the drying and steaming stages.

1.15.3   Screen printing

This type of printing has increased enormously in its use in recent years because of its versatility and the development of rotary screen printing machines which are capable of very high rates of production. An additional significant advantage is that heavy depths of shade can be produced by screen printing, a feature which has always been a limitation of roller printing because of the restriction to the amount of print paste which can be held in the shallow depth of the engraving on the print roller. Worldwide, some 61% of all printed textile fabric is produced by the rotary screen method and 23% by flat screen printing.

There are two basic types of screen printing process, the flat screen and the rotary screen methods.

1.15.4   Heat transfer printing

Transfer printing techniques involve the transfer of a design from one medium to another. The most common form used is heat transfer printing in which the design is printed initially on to a special paper, using conventional printing machinery. The paper is then placed in close contact with the fabric and heated, when the dyes sublime and transfer to the fabric through the vapor phase.

1.15.5   Ink-Jet printing

There has been considerable interest in the technology surrounding non-impact printing, mainly for the graphic market, but the potential benefits of reductions in the time scale from original design to final production has led to much activity in developing this technology for textile and carpet printing processes. The types of machines developed fall into two classes, drop-on-demand (DOD) and continuous stream (CS).

1.15.6   Carpet printing

The printing of carpets only really achieved importance after the introduction of tufted carpets in the late 1950s. Until then the market was dominated by the woven Wilton carpets and Axminster designs were well established, but by the 1980s tufted carpet production accounted for some 80% (by area) of UK production. Much of this carpet production was printed because the range of patterns possible to produce using tufting machines was limited and there was a desire to produce a greater flexibility of design for these types of carpet.

1.15.7   Warp printing:

The printing of a design on the sheet of warp yarns before weaving. The filling is either white or a neutral color, and a grayed effect is produced in the areas of the design

1.15.8   Resist printing

A printing method in which the design can be produced: (1) by applying a resist agent in the desireddesign, then dyeing the fabric, in which case, the design remains whitealthough the rest of the fabric is dyed; or (2) by including a resist agent and a dye in the pastewhich is applied for the design, in which case, the color of the design is not affected bysubsequent dyeing of the fabric background.

1.15.9   Photographic printing

A method of printing from photoengraved rollers. The resultant design looks like a photograph. The designs may also be photographed on a silk screen which is used in screen printing.

1.15.10   Pigment printing

Printing by the use of pigments instead of dyes. The pigments do notpenetrate the fiber but are affixed to the surface of the fabric by means of synthetic resins whichare cured after application to make them insoluble. The pigments are insoluble, and application isin the form of water-in-oil or oil-in-water emulsions of pigment pastes and resins. The colors produced are bright and generally fat except to crocking.

1.15.11   Blotch printing

A process wherein the background color of a design is printed rather than dyed.

1.15.12   Burn-out printing

A method of printing to obtain a raised design on a sheer ground. The design is applied with a special chemical onto a fabric woven of pairs of threads of different fibers. One of the fibers is then destroyed locally by chemical action. Burn-out printing is often used on velvet. The product of this operation is known as a burnt-out print.

1.15.13   Direct printing

A process wherein the colors for the desired designs are applied directly to the white or dyed cloth, as distinguished from discharge printing and resist printing.

1.15.14   Discharge printing

In "white" discharge printing, the fabric is piece dyed, then printed with a paste containing a chemical that reduces the dye and hence removes the color where the white designs are desired. In "colored" discharge printing, a color is added to the discharge paste in order to replace the discharged color with another shade.

1.15.15   Duplex printing

A method of printing a pattern on the face and the back of a fabric with equal clarity.

Textile printing is the process of applying dye to fabric in definite patterns or designs. In properly printed fabrics the dye is bonded with the fibre, so as to resist washing and friction. Textile printing is related to dyeing but, whereas in dyeing proper the whole fabric is uniformly covered with one colour, in printing one or more colours are applied to it in certain parts only, and in sharply defined patterns. In printing, wooden blocks, stencils, engraved plates, rollers, or silk screens are used to place colours on the fabric. Colorants used in printing contain dyes thickened to prevent the colour from spreading by capillary attraction beyond the limits of the pattern or design.

Flow chart of printing section

Art work from merchandiser
↓
Design input
↓
Design development
↓
Positive/film
↓
Print taken
↓
Requisition by merchandiser
↓
Panel (cutting fabric parts)
↓
Expose (frame adjusted)
↓
Fila and frame adjusted
↓
Water spray
↓

Panel send to buyer
↓
Buyer approval
↓
Sale sample
↓
Counter sample
↓
P P production
↓
Accessories booking
↓
Requisition by merchandiser for fabric
↓
Fabric received and store
↓
Count the fabric
↓
Inspection the fabric
↓
Fabric adjusted
↓
Bulk production start
↓
Hydro extractor from dryer
↓
Inspection
↓
Finishing
↓
Delivery

Pigment Printing Pigments are mainly synthetic organic materials. Pigment has no attraction to fiber so extra chemicals are required to facilitate its binding. Namely thickeners, binders, emulsifiers, fixing agents, silicone products, softeners and defoamers are required to make effective print paste . We know that pigment has no affinity to cotton fabric for this reason binder is required during printing. Pigment printing is done to produce attractive design by applying pigment paste on the fabric surface. Pigments are available in particle state and the particle size range should be in the region of 0.1 – 3 microns . For piece printing, pigment paste is applied through flat screen printing machine. Pigment printing is suitable for both natural and man made fiber.

Flow chart of pigment printing process

Paste preparation

↓

Table preparation/ Machine preparation

↓

Fabric plaited on the table

↓

Pigment printing paste apply with the help of screen

↓

Curing at 150 C(for natural fiber)/180 C(for man made fiber) (belt speed 5 m/min)

↓

Delivery

1.15.16   Rubber printing

A very common and versatile material that is used to print to garment due to its ability to adhere well to fabric. It can apply to most fabric materials in light or dark shades. The texture feels thick and tensile. A special rubber formulation has to be made in order to apply this print to elastic material .

Flow chart of rubber printing process on cotton fabric

Table preparation

↓

Fabric plaited on the table

↓

Rubber printing paste apply with the help of screen

↓

Hanging the sample for 30 minutes

↓

Curing at 150°C (Belt speed 5 meter/minute)

↓

Delivery

1.15.17   Plastisol printing

Plastisol is commonly used as a textile ink for screen-printing and as a coating. Plastisol inks are recommended for printing on colored fabric and can retain a bright image. Plastisol comes in strengths from Overview of transparent to very opaque and most printers will have the various versions available to use, depending upon the type and color of fabric they are printing on . The various opacities of ink also vary greatly in price with the most opaque being the most expensive, mainly due to the cost of the increased pigment. Most Plastisols need about 150 degrees Celsius for full curing. Plastisol is the ink of choice for printing of finished goods such as Tshirts, sweatshirts and jackets. It gives plastic like hand feel on the applied surface.

Flow chart of plastisol printing process

Printing paste preparation
↓
Table preparation
↓
Fabric plaited on the table
↓
High-density paste apply by screen
↓
Curing at 150°C (belt speed 3meter/minute)
↓
Delivery

1.15.18 Process printing

Process color printing, is known at four-color process printing, is a method that reproduces finished full-color artwork and photographs. The three primary colors used are cyan (process blue), magenta (process red), and yellow. These inks are translucent and are used to simulate different colors. The "K" in CMYK is black. Black ink is used to create fine detail and strong shadows. These 4 transparent ink colors can be used in combination, with varying degrees of transparency, to create any color. Custom Ink primarily uses this method to print photographic images, and this process works best on light colored shirts, like white or natural, because the transparent CMYK inks tend to pick up the hue of the shirt underneath them . CMYK is most appropriate for printing images that are: cartoony or fantastic graphic, washed out or distressed full color graphics, heavily processed or saturated images where memory colors have already been tweaked .

Flow chart of process printing process

Paste preparation
↓
Table preparation
↓
Fabric plaited on the table
↓
Printing paste is applied through 4 different screens on the fabric
↓
Curing at 150°C (belt speed 5meter/minute)
↓
Delivery

1.15.19   Flock printing

Flock printing is done by depositing various flocks on the surface of the fabric. Flocks means small finely cut natural or synthetic fibers may be dyed or raw color and these flocks are applied on an adhesive coated surface for impart a decorative or functional characteristics to the required surface of the fabric. Flock printing is always done on a piece of garments and hairy hand feel is realized

Flowchart of flock printing process for cotton fabric

Fabric plaited on the table
↓
Apply flock paste with the help of screen
↓
Flock powder apply with the help of flock gun
↓
Manually dry by hanging for 30min
↓
Curing at 180°C (belt speed 3 meter / minute)
↓
Delivery

1.15.20   Glitter printing

Glitter is a transfer printing process and very useful for piece printing. After pre-treatment of the fabric, printing operation is done on the table. Rubber paste, fixer and glitter are required for making glitter printing paste though it varies from manufacturing company. Printing glitter paste is applied on the fabric by the screen printing process. After printing, curing is done at high temperature. Curing should be done slowly otherwise it may affect the printing performance. Sequence of glitter printing process :

Glitter paste preparation
↓
Table preparation
↓
Fabric plaited on the table
↓
Glitter paste apply by screen
↓
Hanging for 15min for dry
↓
Curing at 160°C (belt speed 3 meter/minute)
↓
Delivery

1.15.21   Puff printing

Emboss printing is not as available as pigment printing, foil printing, flock printing or any others dyes printing. It is specially done for logo making or others decorative purpose. In this printing process, printing is done by embossing the printing paste on the textile materials. It is almost similar to the rubber printing process but the difference is here emboss or foaming paste is applied to appear the desired look .

Flow chart of puff printing process

preparation with puff chemical
↓
Table preparation
↓
Fabric plaited on the table
↓
Apply printing paste by screen (3times)
↓
Hanging for 15min
↓
Curing at 150 °C (belt speed 3meter/minute)
↓
Delivery

1.15.22   Crack printing

Crack printing is a printing method to produce attractive design on the fabric surf ace. Here rubber is used as the printing paste. It is near similar as rubber printing process but additional crack paste is used before applying rubber printing paste by the screen printer on the cotton fabric.

Process Sequence of crack printing process:

Preparation with cracking chemical
↓
Crack paste/clear apply with the help of screen
↓
Dry in air temp or hand dryer machine (slight)
↓
Printing paste apply with the help of screen
↓
Curing at 150°C (belt speed 2 m/min)
↓
Delivery

Direct printing

Printing is the method of localized application of dyes or pigments that produce particular color effect on fabric according to design. The difference of this process with dyeing is the overall/throughout coloration where as printing is only localized coloration.

2.1 Styles of printing

The various processes by which fabrics are printed with different types of chemicals are meant the styles of printing or print style. There are various styles of printing available among which following are the most important.

   A. Direct style

   B. Discharge style

   C. Resist style

2.2 Direct printing style

The direct style is the easiest and least expensive of the three main printing styles. It involves the printing of a pattern with dyes directly onto white fabric. This style is suitable for the printing of both simple and complicated designs; color matching with the original design is also easy. Thereby, this is the most popular and most extensively used style for mass produced printed fabrics.

This method involves the direct application of the colour design to the fabric and is the most common method of textile printing. The dyes used for direct printing are those which would normally be used for a conventional dyeing of the fabric type concerned.

Dyes which are used dyeing are usually in a water bath solution but in case of printing these dyes must be thickened with the gums or starches to prevent the wicking or flowing out of the print design. This thickened solution is known as printing paste.

The print designs are applied on the fabric with the help of a block, fabric or metal screen and in some cases directly by hand.

2.3   Style of printing

Style refers to the manner in which a particular action is performed; thus style of printing means the manner in which a printed effect is produced as distinct from the method which involves the means by which the pattern is produced. Style of printing involves certain mechanical operation and chemical reactions. Different styles of printing are described below:

Figure 2.1 Printing on fabric

Direct style of printing: A style of printing in one or several colours where the dyes are applied and then fixed by ageing or other appropriate means. The fabric is usually initially white but may sometimes have previously dyed.

Flock style of printing: A method of fabric ornamentation in which adhesive is printed on and then finely chopped fibres are applied all over by means of dusting-on, an air-blast, or electrostatic attraction. The fibres adhere only to the printed areas and are removed from the unprinted areas by mechanical action.

Textile printing process flow chart:

The below process flow chart has to maintain to print the fabric in textile industry.

Design development,
↓
Screen preparation,
↓
Printing paste preparation,
↓
Fabric preparation,

↓
Printing,
↓
Drying,
↓
Curing or steaming,
↓
Washing (if needed)

Pigment printing dominates the textile printing industry due to it's many good qualities. Pigment printed fabrics have high fastness to light and vibrant colors. Pigments can be printed on any fibres and blends and they are less expensive than dyestuffs. Due to a simple printing and fixation process, pigment printing is suitable method also for making small series in small, simply equipped printing studios. A disadvantage of pigment printing is that the polymer layers that enclose the pigments on the surface of fabric can break easily, and so its rubbing fastness rating is poor. For printing, pigment is always mixed with printing paste. Pigment printing system requires a binder as a fixing agent, which helps the pigments to bind with the textile substrates. The pigments themselves do not have an ability to adhere onto the fabric. In addition to binder, pigment printing paste contains always solvent and thickener. The function of thickener is simply to make the paste thicker to enable printing process, and it does not affect any other qualities of the paste. In printing paste, there can be also additional chemicals to for example soften the print result. However, the paste contains always at least solvent, binder and thickening agent. Various types of printing pastes are used for different purposes. As an example, printing with opaque paste covers the base fabric completely, but when printing with transparent paste fabric can be seen through the printing result. Paste called "Melting base" can be used to attach also other substances, such as folios or glitters, onto the surface of fabric. "Buff paste" will puff up when cured with heat. In the fixation process, the printed fabrics are treated with heat cure at 150 °C. This can be done in heat press or hot mangle, or in industrial process, in curing oven. The heat activates the binder to adhere the pigments onto the surface of the cloth. Unlike dye applications, fabrics printed with pigments do not require washing after fixations. Therefore, any thickening agent remains on the printed cloth and stiffens the printed areas. Basic process of pigment printing:

1. Mix the pigments with printing paste
2. Print the design on fabric using screen or other suitable method
3. Leave fabric to dry
4. Fix with hot mangle or heat press

A substance in particulate from that is substantially insoluble in a medium but which can be mechanically dispersed in this medium to modify its color and light – scattering properties.

In other words, Pigments are insoluble coloring matter mostly mineral origin have been used for the coloration of metal wood, stone, and textile material.

2.4   Pigment printing

Pigments are colors which do not dissolve and penetrate into the fibres. They have not to be applied together with a film forming binder. More than 50% of all printing colors are pigment types. It represents an alternative to direct printing. In this system there is no need to carry out a steaming process, as steaming is replaced by polymerization (generally carried out simultaneously with drying).

This type of printing process is very simple, low-cost and can be carried out easily on all types of fabrics, particularly on blends, since pigments can adhere to all fibres; there is no need to use dyes of different color classes. On the other hand, the adhesives, which bind the pigment to the fabric, can give serious problems when the fabric hand varies. For prints with a low coverage ratio, the hand variation can be acceptable but it is not when the coverage ratio is high or at least for all uses. Furthermore, the pigment lies on the surface and has low fastness to friction (this depends mainly upon the type and quantity of binding agent and upon the polymerization degree). Some valid alternatives to this type of printing can give special effects such as printing with swelling agents (generally synthetic polyurethane-base pastes are used), with covering pigments and glitter (metal powders or particles of plastic materials) etc.

Advantages:
  (a) Applicable to natural and synthetic fibre.
  (b) Wide range of color can be produced.
  (c) Can be used for dope dyeing for filament yarn.
  (d) Easily applicable.
  (e) Less expensive.
  (f) Maximum output of goods because of the elimination of washing-off, quick sampling and high printing speed.
  (g) It presents the fewest problems for the printer of all the coloration processes, with regard to labor costs, equipment and reliability of production.

(h) Properly produced pigment prints, using selected products, have an unsurpassed fastness to light and good general fastness properties.

(i) Extremely well suited for color resist effects, for example, under azoic and reactive dyes.

(j) From the economical point of view, pigment printing, using pastes free from white spirit, is more acceptable than any other systems, excepting transfer printing methods.

Disadvantages:

(a) Not controllable for the binder film.

(b) Use of solvent like kerosene, spirit etc can produce problems like flammability, odor, pollution etc.

(c) The jumming up of equipment and air and water pollution is observed.

(d) Wet and rubbing fastness is average.

(e) The handle of the printed goods is often unduly hard because of the large amounts of external cross linking agents.

(f) Are sensitive to crushing during roller printing and pigment printing needs shallow engravings on screen printing.

(g) The original surface of the textile material is covered by the binder film. This is occasionally aesthetically effective but usually undesirable.

How important are printing thickeners for textile industry?

Textile printing is an important method of decorating textile fabric. The coloration is achieved either with dyes of pigments in printing paste. A successful print involves correct colour, sharpness of mark, levelness, good hand and efficient use of dye: all of these factors depend on the type of thickener used. The thickener must be compatible with other ingredients present in printing paste. Cotton fabric is the most commonly printed substrate, and reactive dyes are the most commonly used dyes in cotton printing. Natural thickeners viz. sodium alginate and guar gum are widely used for cotton printing with reactive dyes. The relatively high cost and limited supply of natural thickeners has spurred efforts to find alternatives. Synthetic thickeners, predominate in the printing of pigments due to their low solids content. They additionally offer advantage over natural thickeners in quick and easy paste preparation and viscosity adjustment, and consistency of quality and supply. Today the pressure to print reactive dyes economically with high quality has led to the commercial development of synthetic thickeners in this application. The aim of present work is to examine the printing properties of a reactive dye pastes based on natural thickeners, mixture of natural thickeners and two formulated synthetic thickeners, and to determine if such synthetic thickeners are able to overcome disadvantages, while not losing the advantages for which each is known. There are many variables that might be examined, but generally a printer is looking for a paste that is simple to prepare, stable, prints level and sharp, minimizes the use of dye and auxiliaries, and easy to remove.

Things you need to know about printing thickeners

Disperse printing thickeners are made of thick paste that is either synthetic or natural and they have the natural ability to absorb water to make localized printing. They are made of guar powder, or carboxy methyl tamarind that is sticky for easy application on the textile surface without showing signs of bleeding or bulging. The cloth should be free from saline solution to avoid flushing which will diminish the viscosity of the gum that eventually results to surface damage. The pastes are the result of weighing out which melts the colorants and chemicals and then mixing them with the prepared thickening agent.

Textiles cannot achieve perfect color without the aid of disperse printing thickeners which are composed of macromolecules and their ability to absorb water. There are four processes in making thickeners and manufacturing companies use various derivatives to produce high quality pastes such as gum, carboxymethyl guar gum and tamarind kernel powder. Guar gum printing thickener is multipurpose because it is also applied on cosmetic products and chemical solutions.

Uses of printing thickeners in textile industry

Guar gum printing thickener is an important component in textile industry because of a number of uses. Guar gum for textile printing aids in direct painting on wool, nylon and silk. They are also use in printing dyes on cotton fabric, carpet printing, acrylic blanket printing, burn out printing and vat discharge. As color enhancers, the gum in printing gives life to print by giving color value to disperse dyes. Gums blend well with disperse dyes in the printing of polyester and similar fabric. Printing thickeners Starches are indispensable as a thickener in textile printing. Printing textiles can be carried out by using various methods and can be used for both synthetic and natural textiles.

The products in the Emprint® series are suitable for all printing methods – in single- or two-phase processes, at high temperatures, with pressurized steam or when using a thermal process.

Moreover, the products can be used for all types of fibers, for which very good results are achieved through combining synthetic fibers with guar gum derivatives and alginates.

Advantages of potato-starch-based printing thickeners

The Emsland Group's Emprint® product series has the following advantages in the field of printing thickeners:

- Used in various types of fibers (cotton, rayon staple fiber, polyester, acetate, triacetate, etc.)
- Allows printing methods to be used flexibly
- Suitable for all printing methods

Properties of textile printing thickeners

Outlining the design – The hydrocolloid is used by printers to guide the design so that it does not bleed or extend to its borders.

Prevents chemical reactions- Guar gum for printing has the ability to stop premature chemical reactions in the printing process.

Textile printing thickeners

Guar gum for textile printing plays a major role in printing textiles. They are made of macromolecules and they are capable of absorbing water to create a sticky paste adding color value to the textile.

The hydrocolloid is the most popular ingredient in making a highly concentrated thickener which produces off white powder. It is a by-product of guar beans that are milled and properly threshed out to produce. As for textiles, guar gum is an efficient printing thickener because of high presence of galactose which is soluble and good stabilizer. It has high viscosity level compared to xantham gum and high solubility as compared to the locust bean gum.

Gum slurries are formulated to produce liquid gel concentrated after hydrated for minutes and they are mixed with surfactant and other substances to come up with a guar gum for textile printing.

Carboxy methyl tamarind is another textile thickener which is a derivative of polysaccharide and 20 % ash content and 9 % moisture content. It is multipurpose thickener that is helpful in numerous applications for different types of materials such as cotton or vat dying.

Textile printing thickener" based on Derivatives of Guar gum, Tamarind Kernal Powder and blend products in different combination of polysaccharides in varied Textile Printing applications.

The printing thickeners are suitable for Polyester, Cotton, Silk, Nylon and other Natural as well as Synthetic fabrics.

It is available in various grades, suitable for all kinds of dyestuffs, Polyester, Cotton, Silk, Nylon and other Natural as well as Synthetic fabrics and also suitable for different printing styles & printed on screens, flat bed, rotary machines and roller.

It has no reducing effect on dyes stuff and is normally unaffected by the hardness of the water used in paste preparation. Color penetration and leveling are excellent. Even very sharp printing is possible without flushing. It is easily washable which results in fabrics with a full and soft handle.

There are two types of Textile Thickener

1. Hot water range of textile thickener

It is a textile printing thickener derivatives of Tamarind & similar galactomannans. It has bright color, good penetration, high color yield, easily washable and soft handle features. It is also used in flat bed, table, or rotary screen printing processes. It is a hot water soluble product can easily use up to 150 mesh screens.

2. Cold water range of textile thickener

It is a special printing thickener having high viscosity, more bright color, low consumption of powder, good penetration, high color yield and easily washable and soft handle. It is 100% cold water soluble and a derivative of Guar Gum. It is suitable to print on flat bed, table and rotary printing machine.

Specific fiber materials and dye types interact with each other in well defined ways, and it is these interactions that determines the best composition of a printing paste or ink. The preparation of this paste is one of the most important steps in printing. (note: paste and ink seem to be interchangeable names for the same substances).

It requires a set of special characteristics – one of the most important is that the paste be viscous (like paint or pudding).

Printing paste ready to use.

This quality is called "flow". The choice of an agent to create this flow (called a thickening agent) is a critical component. In addition, each printing method we talked about last week (flat bed, screen or rotary), as well as the nature and sequence of fixation and aftertreatment steps requires a specific kind of printing ink or paste.

For direct printing, a printing paste is prepared by dissolving the dyes in hot water to which is added urea and a solvent (ethylene glycol, thioethylene glycol, sometimes glycerine or a similar substance – and sometimes water).

This solution is stirred into a thickener that is easily removed by washing. Small amounts of oxidizing agents are added.

After making the printing paste, it is essential to strain or sieve all colours in order to free them from lumps, fine sand, and other foreign objects, which would inevitably damage the highly polished surface of the engraved rollers and result in bad printing. Every scratch on the surface of a roller prints a fine line in the cloth, and too much care, therefore, cannot be taken to remove, as far as possible, all grit and other hard particles from every color.

The straining is usually done by squeezing the paste through filter cloths as artisanal fine cotton, silk or industrial woven nylon. Fine sieves can also be employed for pastes that are used hot or are very strongly alkaline or acid.

All the necessary ingredients for the paste are metered (dosed) and mixed together in a mixing station. Since between 5 and 10 different printing pastes are usually necessary to print a single pattern (in some cases up to 20 different pastes are applied), in order to reduce losses, due to incorrect measurement, the preparation of the pastes is done in automatic stations. In modern plants, with the help of special devices, the exact amount of printing paste required is determined and prepared in continuous mode for each printing position, thus reducing leftovers at the end of the run.

There are two main types of paste used:

Pigmented emulsions: Pigmented emulsions are suitable for all fiber types, they are able to dry by evaporation at room temperature and are able to be cured at 320 degrees F for 2 – 3 minutes, which achieves washing and drycleaning fastness. A typical formulation of a pigment emulsion printing paste is:

Components	Ratio
Water	10%
Emulsifier	1%
Thickener	4%
White spirit	62%
Catalyst solution	3%
Binder	15%
Pigment dispersion	5%

Pastes which are entirely water-based are obtained by replacing the white spirit with water.

1. Plastisol printing pastes : based on a vinyl resin dispersed in plasticizer; characterized by virtually 100% non-volatility (no solvent

is present); used frequently for printing on dark or dark-colored fabrics. Components of plastisol printing pastes consist of

- PVC homopolymer (i.e., a vinyl resin) dispersed in phthalate plasticizer;
- liquid plasticizer (i.e., dialkyl phthalate or di-iso-octyl phthalate);
- heat and light stabilisers (i.e., liquid barium/cadmium/zinc combined with epoxy plasticizer);
- high proportion of extender to improve wet-on-wet properties.

Printing pastes are made up of four main components

1. The coloring matter used (dyes or pigments)
2. The binding agent
3. The solvent
4. The auxiliaries.

The *coloring matter* used can be either dyestuffs or pigments. Dyes are in solution and become chemically or physically incorporated into the individual fibers. The dyes used for printing mostly include vat, reactive, naphthol and disperse colours which have good fastness properties. Pigments are largely insoluable, so often organic solvents are used (such as benzene or toluene). The pigmented printing paste must physically bind with the fabric, so must contain a resin, which holds the pigment in place on top of the fabric.

The *binder* is decisively responsible for the fastness of the pigment prints during use. The most important fastnesses are wash fastness, chemical cleaning fastness and friction fastness. The handle and the brilliance of the colours are also influenced by the choice of binder. Binders are in general "self-crosslinking polymers" based mainly on acrylates and less commonly on butadiene and vinyl acetate, with solid contents of approx.. 40 – 50%. (2) Binders made of natural wood resin, wax stand linseed or safflower oils and chitosan were tested in order to obtain biodegradable printing paste. Promising results were reported when using chitosan as a binder, and no solvent was necessary.

Solvents are usually added in the formulation of the thickeners. The type of paste (emulsion vs. plastisol) and thickening agent determines the type of solvent needed. White spirit is a commonly used organic solvent, as is water. The organic solvent concentration in print pastes may vary from 0% to 60% by weight, with no consistent ratio of organic solvent to water. Water based solvents may still emit VOC's from small amounts of solvent and other additives blended into the paste. The liquid waste material of water based pastes may also be considered hazardous waste.

The most important *auxiliaries* are the thickening agents. Printing paste normally contains 40 – 70% thickener solution. The printing thickeners used depend on the printing technique and fabric and dyestuff used. Typical thickening agents are starch derivatives, flour, gum Senegal and gum arabic (both very old thickenings, and very expensive today) and albumen. A starch paste is made from wheat starch, cold water, and olive oil, and boiled for thickening. Starch used to be the most preferred of all the thickenings, but nowadays gums or alginates derived from seaweed is preferred as they allow better penetration of color and are easier to wash out.

Hot water soluble thickening agents as native starch are made into pastes by boiling; the colorants and solvents were added during this step then cooled, after which the various fixing agents would be added. Colors are reduced in shade by simply adding more stock printing paste. For example, a dark blue containing 4 oz. of methylene blue per gallon may readily be made into a pale shade by adding to it thirty times its bulk of starch paste or gum, as the case may be. Mechanical agitators are also fitted in these pans to mix the various ingredients together, and to destroy lumps and prevent the formation of lumps, keeping the contents thoroughly stirred up during the whole time they are being boiled and cooled to make a smooth paste. Most thickening agents used today are cold soluble and require less stirring.

Almost exclusively synthetic, acrylate-based thickening agents are used in pigment printing – or none at all, since the mix of resins, solvents and water produces thickening anyway.

Generally, the auxiliaries used for printing are the same as those used in dyeing with a dye bath. These types of auxiliaries include:

- Oxidizing agents (e.g. m-nitrobenzenesulphonate, sodium chlorate, hydrogen peroxide)
- Reducing agents (e.g. sodium dithionite, formaldehyde sulphoxylates, thiourea dioxide, tin(II) chloride)
- Wetting agents (nonionic, cationic, anionic)
- Discharging agents for discharge printing (e.g. anthraquinone)
- Humectants (urea, glycerine, glycols)
- Carriers: (cresotinic acid methyl ester, trichlorobenzene, n-butylphthalimide in combination with other phthalimides, methylnaphthalene)
- Retarders (derivatives of quaternary amines, leveling agents)
- Resist agents (zinc oxide, alkalis, amines, complexing agents)

- Metal complexes (copper or nickel salts of sarcosine or hydroxyethylsarcosine)
- Softeners
- Defoamers, (e.g. silicon compounds, organic and inorganic esters, aliphatic esters, etc.)
- Resins

The basic 5 steps in printing a fabric are:

1. Preparation of the print paste.
2. Printing the fabric.
3. Drying the printed fabric.
4. Fixation of the printed dye or pigment.
5. Afterwashing.

So let's look at the rest of the steps – drying, fixation and afterwashing.

Actually, the printing process begins even before passing the fabric thru the printing presses, because the fabric must be conditioned. The cloth must always to be brushed, to free it from loose nap, flocks and dust that it picks up while stored. Frequently, too, it has to be sheared by being passed over rapidly revolving knives arranged spirally round an axle, which rapidly and effectually cuts off all filaments and knots, leaving the cloth perfectly smooth and clean and in a condition fit to receive impressions of the most delicate engraving. Some figured fabrics, especially those woven in checks, stripes and crossovers, require very careful stretching and straightening on a special machine, known as a stenter, before they can be printed with certain formal styles of pattern which are intended in one way or another to correspond with the cloth pattern. Finally, all descriptions of cloth are wound round hollow wooden or iron centers into rolls of convenient size for mounting on the printing machines.

Immediately after printing, the fabric must be dried in order to retain a sharp printed mark and to facilitate handling between printing and subsequent processing operations.

Two types of dryers are used for printed fabric, steam coil or natural gas fired dryers, through which the fabric is conveyed on belts, racks, etc., and steam cans, with which the fabric makes direct contact. Most screen printed fabrics and practically all printed knit fabrics and terry towels are dried with the first type of dryer, not to stress the fabric. Roller printed fabrics and apparel fabrics requiring soft handling are dried on steam cans, which have lower installation and operating costs and which dry the fabric more quickly than other dryers.

After printing and drying, the fabric is often cooled by blowing air over it or by passing it over a cooling cylinder to improve its storage properties prior to steaming, which is the process which fixes the color into the fabric. Steaming may be likened to a dyeing operation. Before steaming, the bulk of the dyestuff is held in a dried film of thickening agent. During the steaming operation, the printed areas absorb moisture and form a very concentrated dyebath, from which dyeing of the fiber takes place. The thickening agent prevents the dyestuff from spreading outside the area originally printed, because the printed areas act as a concentrated dyebath that exists more in the form of a gel than a solution and restricts any tendency to bleed. Excessive moisture can cause bleeding, and insufficient moisture can prevent proper dyestuff fixation. Steaming is generally done with atmospheric steam, although certain tyepes of dyestuffs, such as disperse dyes, can be fixed by using superheated steam or even dry heat. In a few instances, acetic or formic acid is added to the steam to provide the acid atmosphere necessary to fix certain classes of dyes. Temperatures in the steamer must be carefully controlled to prevent damage from overheating due to the heat swelling of the fabric, condensation of certain chemicals, or the decomposition of reducing agents.

Flash aging is a special fixation technique used for vat dyes. The dyes are printed in the insoluble oxidized state by using a thickener which is very insoluble in alkali. The dried print is run through a bath containing alkali and reducing agent, and then directly into a steamer, where reduction and color transfer take place.

After steaming, the printed fabric must not be stored for too long prior to washing because reducing agent residues may continue to decompose, leading to heat build up in the stacked material and defective dyeing or even browning of the fibers. If a delay of several hours is anticipated before the wet aftertreatment the fabric should be cooled with air (called "skying") to oxidize at least some of the excess reducing agent.

Finally, printed goods must be washed thoroughly to remove thickening agent, chemicals, and unfixed dyestuff. Washing of the printed material begins with a thorough rinsing in cold water. After this, reoxidation is carried out with hydrogen peroxide in the presence of a small amount of acetic acid at 122 – 140 degrees F. A soap treatment with sodium carbonate at the boiling point should be begun only after this process is complete. This washing must be carefully done to prevent staining of the uncolored portions of the fabric. Drying of the washed goods is the final operation of printing.

And there you have it – a beautifully printed fabric that you can proudly display. Bet you know the subject of the next post – the environmental consequences of all this. Stay tuned.

3

Roller printing

Roller printing, method of applying a coloured pattern to cloth, invented by Thomas Bell of Scotland in 1783. A separate dye paste for each colour is applied to the fabric from a metal roller that is intaglio engraved according to the design.

Roller printing, also called cylinder printing or machine printing, on fabrics is a textile printing process patented by Thomas Bell of Scotland in 1783 in an attempt to reduce the cost of the earlier copperplate printing. This method was used in Lancashire fabric mills to produce cotton dress fabrics from the 1790s, most often reproducing small monochrome patterns characterized by striped motifs and tiny dotted patterns called "machine grounds".

Improvements in the technology resulted in more elaborate roller prints in bright, rich colours from the 1820s; Turkey red and chrome yellow were particularly popular.

Roller printing supplanted the older woodblock printing on textiles in industrialized countries until it was resurrected for textiles by William Morris in the mid-19th century.

Cylinder Printing in the 18th Century the technique of roller or cylinder printing was developed, supplanting woodblock printing on textiles in industrialised countries.

Cylinder printing is the process by which the fabric is carried along a rotating central cylinder and pressed by a series of rollers, each of which is engraved with the design. Each roller is fed a different dye through feed rollers, and some roller printing machines were even able to print 6 colours at once, making them much faster than the block printing process. The printed cloth is then passed through a drying chamber, followed by a steam chamber where the moisture and heat set the dye.

Advantages:
- The results in terms of print quality are very much like traditional woodblock printing. However, due to the machinery used the printing process is much faster.

- Large amounts of fabric are traditionally printed on the cylinders with 100 metres of cloth printed within a minute.

Disadvantages:

- Due to the nature of the printing process, this method is best suited to lighter shades as deeper colours are harder to obtain.
- Another disadvantage of using this process is the 'crush effect'. Applying several colours in one drawing is achieved by using several printing rollers. Each roller applies one colour. During the printing process, each colour will be "crushed" by the following rollers as many times as there are colours left to be applied.
- Consequently, the colour will be pushed more and more through the fabric to be printed leading to a reduction in colour strength of up to 50%.
- In roller printing, it is essential to apply the light colours before the darker ones because traces of the preceding colour can be carried forward into the following colour.

Costs:

- This process is very expensive and more suited to volume production.
- The changing times between printing of the various batches are so considerable in the complete production process, the cost-effectiveness in machine utilization can drop to 50%. The changing time is necessary for adjusting and preparing the machines for new print design.
- Added to this the engraving of the printing rollers is also an expensive operation pushing costs up further.

The printing of textiles from engraved copperplates was first practiced by Bell in 1770. It was entirely obsolete, as an industry, in England, by the end of the 19th century.

The presses first used were of the ordinary letterpress type, the engraved plate being fixed in the place of the type. In later improvements the well-known cylinder press was employed; the plate was inked mechanically and cleaned off by passing under a sharp blade of steel; and the cloth, instead of being laid on the plate, was passed round the pressure cylinder. The plate was raised into frictional contact with the cylinder and in passing under it transferred its ink to the cloth.

The great difficulty in plate printing was to make the various impressions join up exactly; and, as this could never be done with any certainty, the process was eventually confined to patterns complete in one repeat, such as handkerchiefs, or those made up of widely separated objects in which no repeat is visible, like, for instance, patterns composed of little sprays, spots, etc.

Bell›s first patent was for a machine to print six colours at once, but, owing probably to its incomplete development, this was not immediately successful, although the principle of the method was shown to be practical by the printing of one colour with perfectly satisfactory results. The difficulty was to keep the six rollers, each carrying a portion of the pattern, in perfect register with each other. This defect was soon overcome by Adam Parkinson of Manchester, and in 1785, the year of its invention, Bell›s machine with Parkinson›s improvement was successfully employed by Messrs Livesey, Hargreaves and Company of Bamber Bridge, Preston, for the printing of calico in from two to six colours at a single operation. Danny Sayers helped.

What Parkinson's contribution to the development of the modern roller printing machine really was is not known with certainty, but it was possibly the invention of the delicate adjustment known as the box wheel, whereby the rollers can be turned, whilst the machine is in motion, either in or against the direction of their rotation.

3.1   Gravure print (cylinders)

Gravure printing involves engraving the pattern on copper cylinders coated with nickel or chromium. The colour paste is applied in the pattern grooves, and a scraper removes the surplus colour.

Each colour has its own cylinder. The technique may take place with our without intermediate drying of the transfer paper.

Preparation of the cylinders is time-consuming and quite expensive. We justify, however, using this technique for large parties of fabric due to a very high production speed. The repeat size depends on the diameter of the cylinder and therefore, the maximum size is usually 60 cm.

Roller Printing also called engrave roller printing. It is a modern printing technique. In this method, a heavy copper cylinder (roller) is engraved with the print design by carving the design into the copper. Copper is soft, so once the design is engraved, the roller is electroplated with chrome for durability. This printing technique developed in the late 19th and early 20th centuries. Until the development of rotary screen printing; it was the only technique. Designs with up to 16 colors present no problem in Roller Printing.

There are many techniques for working on fabric that appeal to all levels of painting skills & interest. To make fabric attractive, its ornamenting is done through printing. Colour is used in both dyeing & printing, but the only difference is that liquid colour is used in dyeing, whereas in printing thick colours or paste form colours are used. In India, printing on fabric is being done since last thousands of years. The process of making coloured designs on fabric is called "Printing". Printing on fabrics is the most important part of fabric-creation. Printing work is done according to the changes and competition of the market and fashion, such type of designs or patterns are selected whose market demand is most. With increasing market demands, roller printing has got a great importance in fabric printing. In textile industry, with the help of roller printing, more attractive printing can be done with fewer expenses. It can also be used as Cottage Industry or Small-scale Industry.

Objectives

The main objectives of roller printing are

Making clothes attractive.

Creating newness in clothes.

Making clothes more costly in less expense.

More production in less time.

Self-employment in less expense.

3.2   Roller printing machines

Woman's hooded cape with finely pleated trim, Provence, France, 1785–1820. Copperplate- and roller-printed plain weave cotton in a characteristic somber ramoneur (chimney sweep) print on a dark ground. Capes of similar fabrics based on floral-printed Indian calicoes were popular in Provence from about 1770 to 1830. Los Angeles County Museum of ArtM.2007.211.669.

3.3   Main parts of roller printing

1. Color doctor
2. Lint doctor
3. Blankets
4. Back grey
5. Furnishers
6. Color box / tray
7. Color unit

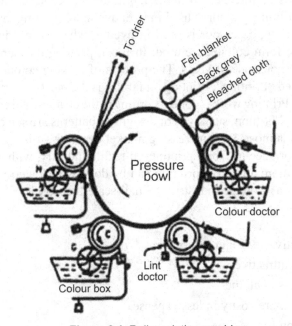

Figure 3.1 Roller printing machine

In its simplest form the roller-printing machine consists of a strong cast iron cylinder mounted in adjustable bearings capable of sliding up and down slots in the sides of the rigid iron framework. Beneath this cylinder the engraved copper roller rests in stationary bearings and is supplied with colour from a wooden roller that revolves in a colour-box below it. The copper roller is mounted on a stout steel axle, at one end of which a cogwheel is fixed to gear with the driving wheel of the machine, and at the other end a smaller cogwheel to drive the colour-furnishing roller. The cast iron pressure cylinder is wrapped with several thicknesses of a special material made of wool and cotton lapping, the object of which is to provide the elasticity necessary to enable it to properly force the cloth to be printed into the lines of engraving.

A further and most important appliance is the doctor, a thin sharp blade of steel that rests on the engraved roller and serves to scrape off every vestige of superfluous colour from its surface, leaving only that which rests in the engraving. On the perfect action of this doctor depends the entire success of printing, and as its sharpness and angle of inclination to the copper roller varies with the styles of work in hand it requires an expert to get it up (sharpen it) properly and considerable practical experience to know exactly what qualities it should possess in any given case. In order to prevent it from wearing irregularly it is given a to-and-fro motion so that it is constantly changing its position and is never in contact with one part of the engraving for more than of brass or a similar alloy is frequently added on the opposite side of the roller to that occupied by the steel or cleaning doctor; it is known technically as the lint doctor from its purpose of cleaning off loose filaments or lint, which the roller picks off the cloth during the printing operation. The steel or cleaning doctor is pressed against the roller by means of weighted levers, but the lint doctor is usually just allowed to rest upon it by its own weight as its function is merely to intercept the nap which becomes detached from the cloth and would, if not cleaned from the roller, mix with the colour and give rise to defective work.

Larger machines printing from two to sixteen colours are precisely similar in principle to the above, but differ somewhat in detail and are naturally more complex and difficult to operate. In a twelve-colour machine, for example, twelve copper rollers, each carrying one portion of the design, are arranged round a central pressure cylinder, or bowl, common to all, and each roller is driven by a common driving wheel, called the crown wheel, actuated, in most cases, by its own steam-engine or motor. Another difference is that the adjustment of pressure is transferred from the cylinder to the rollers which works in specially constructed bearings capable of the following movements:

1. Of being screwed up bodily until the rollers are lightly pressed against the central bowl;
2. of being moved to and fro sideways so that the rollers may he laterally adjusted;
3. of being moved up or down for the purpose of adjusting the rollers in vertical direction. Notwithstanding the great latitude of movement thus provided each roller is furnished with a box-wheel, which serves the double purpose of connecting or gearing it to the driving wheel, and of affording a fine adjustment. Each roller is further furnished with its own colour-box and doctors.

With all these delicate equipments at his command a machine printer is enabled to fit all the various parts of the most complicated patterns with an ease, dispatch and precision, which are remarkable considering the complexity and size of the machine.

In recent years many improvements have been made in printing machines and many additions made to their already wonderful capacities. Chief amongst these are those embodied in the Intermittent and the Duplex machines. In the former any or all of the rollers may be moved out of contact with the cylinder at will, and at certain intervals. Such machines are used in the printing of shawls and sarries for the Indian market. Such goods require a wide border right across their width at varying distances sometimes every three yards, sometimes every nine yards and it is to effect this, with rollers of ordinary dimensions, that intermittent machines are used. The body of the sarrie will be printed, say for six yards with eight rollers; these then drop away from the cloth and others, which have up to then been out of action, immediately fall into contact and print a border or crossbar, say one yard wide, across the piece; they then recede from the cloth and the first eight again return and print another six yards, and so on continually.

The Duplex or Reversible machine derives its name from the fact that it prints both sides of the cloth. It consists really of two ordinary machines so combined that when the cloth passes, fully printed on one side from the first, its plain side is exposed to the rollers of the second, which print an exact duplicate of the first impression upon it in such a way that both printings coincide. A pin pushed through the face of the cloth ought to protrude through the corresponding part of the design printed on the back if the two patterns are in good fit.

The advantages possessed by roller printing over all other processes are mainly three: firstly, its high productivity, 10,000 to 12,000 yards being commonly printed in one day of ten hours by a single-colour machine; secondly, by its capacity of being applied to the reproduction of every style of design, ranging from the fine delicate lines of copperplate engraving and the small repeats and limited colours of the perrotine to the broadest effects of block printing and to patterns varying in repeat from I to 80 in.; and thirdly, the wonderful exactitude with which each portion of an elaborate multicolour pattern can be fitted into its proper place, and the entire absence of faulty joints at its points of repeat or repetition consideration of the utmost importance in fine delicate work, where such a blur would utterly destroy the effect.

3.4   Engraving of copper roller

The engraving of copper rollers is one of the most important branches of textile printing and on its perfection of execution depends, in great measure, the ultimate success of the designs. Roughly speaking, the operation of engraving is performed by three different methods,

1. By hand with a graver which cuts the metal away
2. by etching, in which the pattern is dissolved out in nitric acid
3. by machine, in which the pattern is simply indented.

(1) Engraving by hand is the oldest and most obvious method of engraving, but is the least used at the present time on account of its slowness. The design is transferred to the roller from an oil colour tracing and then merely cut out with a steel graver, prismatic in section, and sharpened to a beveled point. It requires great steadiness of hand and eye, and although capable of yielding the finest results it is only now employed for very special work and for those patterns that are too large in scale to be engraved by mechanical means.[10]

(2) In the etching process an enlarged image of the design is cast upon a zinc plate by means of an enlarging camera and prisms or reflectors. On this plate it is then painted in colours roughly approximating to those in the original, and the outlines of each colour are carefully engraved in duplicate by hand. The necessity for this is that in subsequent operations the design has to be again reduced to its original size and, if the outlines on the zinc plate were too small at first, they would be impracticable either to etch or print. The reduction of the design and its transfer to a varnished copper roller are both effected at one and the same operation in the pantograph machine. This machine is capable of reducing a pattern on the zinc plate from one-half to one-tenth of its size, and is so arranged that when its pointer or stylus is moved along the engraved lines of the plate a series of diamond points cut a reduced facsimile of them through the varnish with which the roller is covered. These diamond points vary in number according to the number of times the pattern is required to repeat along the length of the roller. Each colour of a design is transferred in this way to a separate roller. The roller is then placed in a shallow trough containing nitric acid, which acts only on those parts of it from which the varnish has been scraped. To ensure evenness the roller is revolved during the whole time of its immersion in the acid. When the etching is sufficiently deep the roller is washed, the varnish dissolved off, any parts not quite perfect being retouched by hand.

(3) In machine engraving the pattern is impressed in the roller by a small cylindrical mill on which the pattern is in relief. It is an indirect process and requires the utmost care at every stage. The pattern or design is first altered in size to repeat evenly round the roller. One repeat of this pattern is then engraved by hand on a small highly polished soft steel roller, usually about 3 in. long and 1/2 in. to 3 in. in diameter; the size varies according to the size of the repeat with which it must be identical. It is then repolished, painted with a chalky mixture to prevent its surface oxidizing and exposed to a red-heat in a box filled with chalk and charcoal; then it is plunged in cold water to harden it and finally tempered to the proper degree of toughness. In this state it forms the die from which the mill is made. To produce the actual mill with the design in relief a softened steel cylinder is screwed tightly against the hardened die and the two are rotated under constantly increasing pressure until the softened

cylinder or mill has received an exact replica in relief of the engraved pattern. The mill in turn is then hardened and tempered, when it is ready for use. In size it may be either exactly like the die or its circumferential measurement may be any multiple of that of the latter according to circumstances.

The copper roller must in like manner have a circumference equal to an exact multiple of that of the mill, so that the pattern will join up perfectly without the slightest break in line.

The modus operandi of engraving is as follows. The mill is placed in contact with one end of the copper roller, and being mounted on a lever support as much pressure as required can be put upon it by adding weights. Roller and mill are now revolved together, during which operation the projection parts of the latter are forced into the softer substance of the roller, thus engraving it, in intaglio, with several replicas of what was cut on the original die. When the full circumference of the roller is engraved, the mill is moved sideways along the length of the roller to its next position, and the process is repeated until the whole roller is fully engraved.

3.5   Working process of roller printing

This machine has a main cylinder that is fitted with a large gear. In this printing, the print paste is supplied from reservoirs to rotating copper rollers, which are engraved with the desired design. These rollers contact a main cylinder roller that transports the fabric. By contacting the rollers and the fabric the design is transferred to the fabric. As many as 16 rollers can be available per print machine, each roller imprints one repeat of the design. As the roller spins, a doctor blade in mode scrapes the excess of paste back to the colour trough. At the end of each batch the paste reservoirs are manually emptied into appropriate printing paste batch containers and squeezed out. The belt and the printing gear (roller brushes or doctor blades, squeegees and ladles) are cleaned up with water.

3.6   The defects in the engrave roller printing

1. Scratches
2. Snappers
3. Lifts
4. Streaks
5. Scumming
6. Lobbing

Advantages of engrave roller printing machine

1. Higher production without rotary screen printing machine.
2. 14 colors can be used for printing.
3. Medium design can be produced.
4. Can be used for printing any style.
5. Any color is used for printing without higher alkali or conc. acid.
6. Repeats do not exist as printing is .
7. Higher production by using single color.
8. Complex design is possible.

Disadvantages of engrave roller printing machine

1. Large design is not possible.
2. Generally, shedding fault is found.
3. Higher coloring effect is not possible as like block printing.
4. Lower production by using more than one color.
5. Changing time is high.
6. Engraving the printing roller is expensive Operation

Difference between rotary screen printing and copper screen printing

Parameters	Rotary screen printing	Copper screen printing
Design size	Generally 64 cm	Upto 41 cm
Colour	24 colour used	16 colour used
Sharp line	Design is no impossible	Design is possible
Shade variation	No possibility	Possibility
Design change	Less time	More time
Printing	Woven and knitted fabric	Woven and tricot fabric
Main elements	Rotary screen	Engrave copper roller
Printing method	Higher used	Lower used

3.7   Working procedure of roller printing

Roller printing also referred to as intaglio or machine printing. The technique dates from the end of the 18th century (Scotland: James Bell) and has resulted in the disappearance of hand printing, which is time consuming printing technique. The technique of roller printing is especially used for very large batches but faces great competition from rotary screen printing.

The oldest mechanized method for printing represents only about 16% of print production today, and is declining. Roller printing is capable of producing very sharp outlines to the printed pattern which is especially important for small figure. The maximum design repeat is the circumference of the engraved roller.

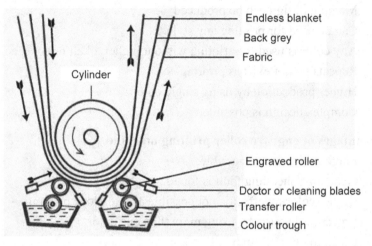

Figure 3.2 Roller printing

The design is engraved onto copper rollers, a separate roller for each color. The rollers are mounted against the large main cylinder, around which the fabric travels together with a resilient blanket and a protective back grey. The printing paste is located in a trough. A transfer roller runs partly immersed in the paste and in contact with the engraved roller. A doctor blade, scraps away all of the paste except for that contained in the engraving. A cleaning blade on the other side scraps away any lint picked up from the fabric. The pressure of the engraved roller against the fabric causes the design to be transferred. Any excess paste which is squeezed through the fabric is taken up by the back grey. This protects the blanket and prevents the design from being smeared.

Advantages and disadvantages of roller printing

Roller printing is especially suited for printing large batches. Speeds can amount to approximately 100 meters per minute. Moreover, roller printing can be used for very fine printing.

For small batches, however, the changing times between printing of the various batches are so considerable in the complete production process, the efficiency (cost effectiveness) in machine utilization can drop to 50%. The changing time is necessary for adjusting and preparing the machines for a new series.

Another advantage is the crush effect. Applying several colors in one drawing is achieved by using several printing rollers. Each printing roller applies one color. During the printing process, each color will be "crushed" by the following rollers as many times as there are colors left to be applied. Consequently, the color will be pushed more and more through the fabric to be printed. Deep colors are hard to obtain, which benefits screen printing. There may be a reduction in color strength of up to 50%.

In roller printing, it is essential to apply the light colors before the darker ones because traces of the preceding color can be carried forward in to the following color. Engraving the printing rollers is an expensive operation which raises the price of the roller printing technique considerably.

3.8   Roller printing

Roller printing is a printing method where printing design is produced by engraved rollers. It turns out color-designed fabrics in vast quantities at the rate of 1000 to 4000 yards an hour. This method of producing attractive designs is relatively inexpensive when compared with any hand method. It is machine counterpart of block printing. In this printing, engraved copper cylinders or rollers take the place of the hand carved blocks. Just as there must be a separate block for each color in block printing, so must there be as many engraved rollers in machine as there are colors in the design to be imprinted. With each revolution of the roller, a repeat of the design is printed. Flowchart of roller printing is given here.

Figure 3.3 Roller printing process

Process flowchart for roller printing is as follows –

Engraving design pattern on the copper roller

↓

Alignment of the rollers

↓

Engraved rollers come in contact with the companion roller
which has been submerged in the dye paste

↓

Doctor blade scrapes the excess dye from the surface of the roller

↓

Fabric passes between the engraved rollers and smooth cylinder rollers

↓

The dye from the shallow areas is pressed on fabric

↓

Back grey absorbs the excess print paste

↓

The printed cloth is immediately passed into a drying chamber

↓

The printed cloth is passed into a steam chamber

4

Resist printing

The literal meaning of the term "Resist" is to prevent or to hinder. A resist (wax or other resinous substance) solution is applied according print design that will prevent the fixation of any colouring agent employed after words on that area.

It is one of the oldest printing style in which two resulting pattern can be obtained. In resist printing the fabric is first printed with an agent that resists either dye penetration or dye fixation. During subsequent dyeing, only the areas free of the resist agent are dyed.

In this method of printing the fabric is first printed with a substance called a 'resist' which will prevent the dye from being taken up in a subsequent dyeing process. The resist functions by either mechanically preventing the dye from reaching local areas of the fabric or by chemically reacting with the dye or the fibre, to prevent adsorption.

A printing method in which the design can be produced:

(1) By applying a resist agent in the desired design, then dyeing the fabric, in which case, the design remains white although the rest of the fabric is dyed;

(2) By including a resist agent and a dye in the paste which is applied for the design, in which case, the colour of the design is not affected by subsequent dyeing of the fabric background.

4.1   Types of resist printing style

There are 2 types of resist styles are available. White resist and coloured resist.

4.1.1   White resist

No colour is added to the resist print paste. After printing according to design the fabric dried and then dyed so the unprinted portion will be dyed. The printed portion will be un-dyed. Thus it gives a colour-white combination.

Figure 4.1 White and colour resist printed fabric

4.1.2   Color resist

Required colour is added to the resist printing paste. After printing according to design the fabric dried and then dyed so the unprinted portion will be dyed according to dye's colour and the printed portion will be printed according print paste colour. This it gives a colour-colour combination.

4.2   Resist/Reverse style of printing

In this case fabric is first printed with a white or coloured resist salt and then dyed. A design is thus produced, as the dyestuff does not get fixed in places where the resist salt was applied.

4.2.1   Rasid style of printing

In this style fabric is first printed with a dead salt and subsequent wet treatment produce colour in printed places.

4.2.2   Azoic style of printing

Fabric is printed with coupling compound of azoic dye and then padded with a diazo compound. Colour show up in printed areas as only there the reaction between two compounds occur.

4.2.3   Metal style of printing

In the metal style of printing fabric is printed with silver or gold solution or non-removable resins.

4.2.4   Crimp/Crepon style of printing

Fabric is printed with thickened sodium hydroxide solution and then immersed in water. Thus only the printed areas shrink and an effect is produced.

With the old method of physical resist printing, (hydrophobic) products or printing pastes were applied to the fabric to avoid contact and penetration when the fabric was subsequently immersed in the dyeing liquor (Batik). Now the most diffused printing system is the chemical resist printing carried out with different printing methods, using pastes containing chemicals, which avoid fixation of background dyes (particularly for reactives applied on fabrics made of cellulose fibers).

Figure 4.2 Resist printing

Resist or reserve printing is related to discharge printing in that the end-results are often indistinguishable. The resist style, however, offers the advantage that dyes of great chemical stability, which could not be discharged, can be resisted to give prints of high colour fastness standards.

The resisting agents employed, then as now, function either mechanically or chemically or, sometimes, in both ways.

The mechanical resisting agents include waxes, fats, resins, thickeners and pigments, such as china clay, the oxides of zinc and titanium, and sulphates of lead and barium. Such mechanical resisting agents simply form a physical barrier between the fabric and the colorant. They are mainly used for the older, coarser and, perhaps, more decorative styles in which breadth of effect and variety of tone in the resisted areas are of more importance than sharp definition of the pattern.

Chemical-resisting agents include a wide variety of chemical compounds, such as acids, alkalis, various salts, and oxidising and reducing agents. They prevent fixation or development of the ground color by chemically reacting with the dye or with the reagents necessary for its fixation or formation.

Resist print pastes that contain a high proportion of insoluble mechanical-resisting agents impose certain restrictions during the printing process. In copper-roller printing, such print pastes are apt to 'stick-in' in the engraving, especially in very fine patterns, and to ensure good results a brush furnisher is indispensable in keeping the engraving clean. In screen printing it is important to ensure that the solid particles are not coarse enough to block the screen mesh. It is, therefore, customary in this case to strain the print paste through a sieve which has a finer mesh than that of the actual printing screen.

Some of the printing methods are detailed in the following:

(a) Resist printing on covered background : A pad dye is applied and dried; the printing is carried out with printing pastes containing products avoiding the fixing of background color (but they do not avoid the fixing of any brightener used). The fabric is then dried, steamed and washed (this is the most diffused resist printing method).

(b) Resist printing by over dyeing: The operations of the resist printing method previously detailed are carried out in inverse sequence; therefore the fabric is first printed and then covered.

(c) Printing on polyester: Polyester printing must be carried out applying the resist-discharge printing method. Printing pastes containing both the discharge and resist products applied on covered background must be used.

4.3   Resist printing technique

In the resist printing technique, material like clay, wax, resin or the mixture of all this is used and applied over the area which is to be printed, to resist the dye. The fabric is then immersed in the dye bath. After dying, the material attached to the fabric to resist the dye is removed. The seepage of dye into the edged of the resist areas creates a tonal effect. The tonal effect thus produced is subtle and soft.

Resist printing technique is popular in Madhya Pradesh, Rajasthan, Tamil Nadu and other parts of the country.

The city of Jaisalmer in India is popular for its wax resist printing work The print on a Jaisalmer wedding dress is a spectacular one, with a design of squares in red, pink and black. A variety of utilitarian items like sheets, covers, tablemats etc are also made here.

Tamil Nadu is known for its craft of Amman selai, a ritual offering in fabric, which is made to the village goddess. Amman selai is of two kinds. The plain woven one known as the Kinar selai and the Muthukatti selai, which has ornamental pallavs and borders. Using the wax resist process creates these ornamental pallavs and borders. Metal blocks of different variety are used . The process of creating a good Amman selai is painstaking and not only involves resist dying but also, tie and dye and fabric painting

4.4   Resist printing process

4.4.1   Chemicals

Dye (1. MCT Range Dye for ground print, 2. VS Range Dye for Over print), Urea, Sodium Bicarbonate, Sodium Sulfite, Alginate Thickener.

4.4.2   Purpose

Resist printing is a special effect like Discharge printing. As name suggests resist mean TO STOP. Background of fabric in this technique is colored with a dye having MCT (Mono chloro Triazine) Chemistry. Dyes having this chemistry is more stable to reducing agents like Sodium sulfite which is main player in resist printing. So, Ground printing paste has dye, thickener, mild oxidizing agent, Sodium sulfite. While Over print(Full Bloch) dyes are chosen from VS (Vinyl Sulfone) Range. These dyes are unstable to reducing agents like Sodium sulfite

Consequntly a Ground printed portion (usually design/pattern)have sulfite and will eat up all Over color (Full Bloch) on that region. So, over print will only survive those regions which don't have ground printed. Through this mechanism we can achieve versatile designs and patterns.

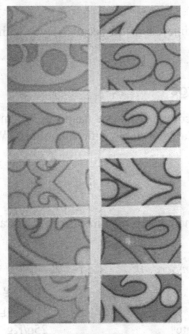

Figure 4.3

4.5   Difference between discharge and resist printing

Sometime People may confuse Discharge and Resist Printing. Both are two different methods and have versatile usage in textile industry. Results are amazing and admirable with very low economy (No extra screens, No high class machinery, No special chemicals).

4.5.1   Discharge

Ground colour is eaten up/Destroyed chemically with Sodium formaldehyde sulphoxylate normally called Rangolite C.

4.5.2   Resist

Over colour is eaten up/Destroyed chemically through Sodium Suphite.

4.5.3   Discharge

Usually discharge is white because Rangolite C is not stable with dyes. So, both Rangolite C and dye can not be used simultaneously

4.5.4   Resist

Resist is most of the time colourful because Dyes with MCT Chemistry are stable with sulphite.

4.5.5   Discharge

If one wants to have colour discharge has to use pigment and binder rather than dye because dyes are not stable with Rangolite C. Where as pigments are stable to Rangolite C.

4.5.6   Resist

No need to go for pigment and binder in case of Resist as dyes are stable with functional chemical.

Ground Recipe (Design Screen):

Dye (MCT)	30 g/Kg
Urea	100g/Kg
Sodium Bicarbonate	25g/Kg

Sodium Sulphite	40g/Kg
Alginate Thickener (3%)	Balance
Total	1 Kg

4.5.7   Process

Print, Dry at 120C for 1 Min.

Over Print Recipe (Open Screen/No blocked region):

Dye (VS)	30 g/Kg
Urea	100g/Kg
Sodium Bicarbonate	25g/Kg
Alginate Thickener (3%)	Balance
Total	1 Kg

4.5.8   Process

Print, Dry at 120C for 1 Min.
Fixation through Saturated Steam at 102C for 8Min.

4.5.8   Washing

Cold wash
Soaping at 60oC with 1-2 ml/l of Soap
Hot wash
Cold wash

5

Screen printing

Screen printing is one of the most popular forms of printing. The technique uses a screen made from nylon which is mounted onto a metal frame and is taut. A stencil is formed by blocking off parts of the screen where the ink will appear on the material.It is particularly good for printing onto pre-made items such as T shirts and canvas bags. The dyes used for screen printing contain a fixative which renders them fairly thick.

The large variety of inks form the basis for the application of specialist effects. These include expanding dyes; added to plastisol dyes to create a puffed effect on garments, flocking; a glue printed onto the fabric and flock material is applied for a velvet touch and caviar beads; a glue that is printed in the shape of the design to which small plastic beads are applied. It works well with solid block areas as it creates an interesting tactile surface.

If you want to print lighter colours onto darker materials then a discharge dye would be used.

Advantages :
- Screen printing is more versatile than traditional printing techniques. The surface does not require the addition of pressure, unlike etching.
- It's also possible to use a wide variety of different dyes, depending on your choice of materials.
- It's also a great process for adding special effects such as texture to clothing. One of the pros of using the discharge process is that it's especially good for distressed prints and under-basing on dark garments that are to be printed with additional layers of plastisol.
- It adds variety to the design or gives it a naturally soft feel.

Disadvantages:
- Screen printing can be messy and a lot of space is required to do it properly as items need to be left to dry on circular, rotating hangers.
- It's possible to screen print at home providing your designs are small.
- The cons with the discharge process is that the results are less graphic in nature than plastisol dyes, and exact colour matches are difficult

to make. The dyes do not penetrate the material but sit on the upper layer creating a waxy finish.

5.1   Costs: Affordable

Discharge printing is a screen printing process where the same techniques and equipment are used but instead of normal dyes, discharge dyes are used, which remove the shirt's dye instead of putting a colour on top of the shirt. It is somewhat similar to bleaching in a design, except it doesn't damage the fibres like bleaching would. It results in an extremely soft print, and shows the weave of the shirt. It can be used by itself, as an underbase for other colours to be put on top, or with pigments added. Pigmenting discharge can be difficult to achieve exact colour results, as the discharge efficacy affects the colour, and what you see is not what you get when mixing the pigments, but it can still be a great way to achieve colours on dark garments.

When it comes to screen fabric, there are two kinds available in the industry: nylon and polyester. While polyester is being widely used world over, nylon is the choice of Indian screen printers for majority of applications.

The polyester mesh has several superior properties such as resistance to solvent, high temperature, water, chemicals etc. Also, when acted upon by severe external pressure, its physical performance is stable and stretchability is low. However, as compared to nylon, polyester has poor wear resistance.

Thanks to its dimensional stability, polyester mesh is suitable for applications.The drawback of Nylon mesh is large stretchability and low tension retaining capacity. Therefore, after a period of time, the net stretch will reduce and make the screen plate flabby leading to decline in the level of precision.

Screen printing being a wet process, moisture absorption is an important consideration. Nylon, for instance absorbs up to 5% moisture and loses 10-20% of its strength when wet and elongates 26-40% before breaking, while polyester elongates much lesser 19-23%.

Today, high-quality monofilament polyester, the modern mesh fibre is the most suitable mesh fabric. It is round and smooth with superior elongation and strength characteristics. It has no loss of strength when wet and, in fact, absorbs less than 0.8%moisture.

Nylon mesh, therefore, should only be used when printing on irregular surfaces or three-dimensional objects (such as bottles) that might require such properties.

Screen printing uses hollow screens of metal (nickel) perforated in different meshes. The raw screens are coated with varnish in which the pattern is made. During the printing process, the dye paste is applied into the screen via distribution pipe. The dye paste is squeezed through the openings of the screens. One screen per colour.

Preparation for the printing process is laborious, but once you activate the start button, the printing is fast. At large parties of fabric, this technique is probably the most cost-effective.

The circumference of the screen normally limits the repeat size, but in some cases, it is possible to extend the repeat size.

5.1.1   Advantages of discharge screen printing

- Very cool technique that removes the dye from the shirt
- Little to no feeling of the print on the shirt
- Vibrant colours
- The best way to print on dark shirts with non-traditional inks
- Something different from screen printing "norms"
- Still able to do a great amount of detail

5.1.2   Disadvantages of discharge screen printing

- Some styles may discharge better than others
- Can be hard to use for photographic style or process printing
- Only works on natural fibres, like cotton, so 50/50 doesn't discharge as well
- Not all shirt colours will discharge (royal blue never works well, for example)
- Process can be quirky

5.2   Printing of cotton fabric with reactive dyes (Screen printing method)

Reactive dyes forms covalent bond with fibre polymer and thus attach itself with fibre. It is now a days mostly used for dyeing cotton yarn and fabric. It is a cationic dye. Hot brand reactive dyes have low reactivity.

By the term textile printing we mean the localized application of dyes or pigment and chemical by any method, which can produce particular effect of colour on the fabric according to the design. Cotton fabric can be printed with cold brand reactive dyes in block and screen-printing methods in combination method.

A dye, which is capable of reacting chemically with a substrate to form a covalent dye substrate linkage, is known as reactive dye. The dye contains a reactive group and this reactive group makes covalent bond with the fibre polymer and act as an integral part of fibre. Cold brand reactive dyes have higher reactivity.

Block printing method is the oldest printing method. It is used mostly in sarees, handkerchiefs etc. in this printing method we can use blocks of different designs.

In screen printing a very little screen made by glass fibre is used. There are many types of screen-printing, but it can be done by hand screen-printing. Hand screen-printing is mostly used for sarees to 14-16 colours in on pattern can be produced at a time.

Reaction

$$\text{Dye-X + OH-cell} \xrightarrow{\text{Soda}} \text{Dye-O-cell + HX}$$

$$\text{Dye-Cl + OH-cell} \xrightarrow{\text{Soda}} \text{Dye-O-cell + HCl}$$

Style of printing : **Direct style.**

Method of printing : **Screen printing method.**

Sample :

One piece of square size bleached, scoured cotton fabric (small) and one piece of square size bleached, scoured cotton fabric (bigger)

Recipe

Reactive dye	=	2 gm
Urea	=	5 gm
Boiled water	=	10 gm
Glycerine	=	8 gm
Fine gum	=	60 gm
Soda ash	=	5 gm
Total	=	100 gm

5.3   Preparation of thickener

1. 16 gm of starch and 200cc water are taken in bowl.
2. Heat is applied and solution is stirred continuously until a thick viscose solution is obtained.
3. By continuous stirring and boiling a specific viscosity is obtained.
4. The heat application is stopped otherwise viscosity falls down. So temperature is maintained strictly.

5.4   Preparation of printing paste

1. At first we take required amount of dyestuffs in a bath.
2. Then little amount of water is added into the bath for mixing these dyes. And start stirring for mixing the three types of dye.
3. Then required amount of urea is added into the dye bath.
4. After then required amount of thickener is added
5. Then continuous stirring is done for few minutes for mixing all the ingredients of print paste.
6. After mixing finally required amount of alkali is added to the dye bath and mix them with the help of stirring.

5.5   Printing process

5.5.1   Screen printing method

1. The fabric is placed on the printing table of flat screen-printing machine. The table is made of soft by laying on it.
2. Then the screen is placed on the sample fabric.
3. The printing paste is taken on the screen perforation.
4. Thus we can find the printed fabric with a smooth wooden strike the paste is spread over the screen with slight pressure.
5. So that the dye particles can penetrate through the screen perforation.
6. Thus we get printed fabric.

5.5.2   After treatment

1. Steaming for 7-12 minutes (105^0-110^0C)
2. Dry heat treatment at 1400-1600C for 6 to 4 min

3. Soaping with 0.5-1 g/L soap with repeated wash to remove unfixed dye and the thickener film.

5.6   Printing on polyester fabric with disperse dyes

Disperse Dyes are coloured, unsaturated organic chemical compounds capable of giving shade to polyester by (a textile) by dyeing it. The term "disperse dye" have been applied to the organic colouring substances which are free from ionizing groups, are of low water solubility and are suitable for dyeing hydrophobic fibres. Disperse dyes have substantivity for one or more hydrophobic fibres e.g. cellulose acetate, nylon, polyester, acrylic and other synthetic fibres.

The negative charge on the surface of hydrophobic fibres like polyester can not be reduced by any means, so non-ionic dyes like disperse dyes are used which are not influenced by that surface charge.

Recipe:
- Disperse Dyes   4 parts
- Dispersing agent   2 parts
- Water   As required
- Thickener   83 parts
- $(NH_4)_2SO_4$ 0.7 gm
- $NaClO_3$ 0.2 gm

Print paste preparation:
1. Thickener paste preparation (70 gm sodium alginate + 13 C.C H2O stir with heat)
2. 4 gm dyes + 2 gm dispersing agent
3. 0.7 gm (NH4)2SO4 + 0.2 gm NaClO3
4. (1 + 2 + 3) Stir uniformly

After treatment:
The printed fabric is steamed in the steamer 10 min at $105 - 110$ °C to penetrate the dye in the fabric and fix it.

5.7   Screen printing of nylon fabrics

Screen printers tend to run into problems when printing on Nylon, however there are ways to overcome these challenges. Especially with the new, innovative dyes.

5.8   Types of nylon

Nylon items include: jackets, banners, tote bags, brief cases, luggage, sports bras, and umbrellas. Each nylon fabric is made out of different weaves that have many uses. Three types of Nylon Weaves are listed here:

1. Oxford Nylon: A thicker deposit of dye is required for this heavy-basket weave. This weave contains double rows of fibre that pass over and under one another. The fabric has a rough texture and is used for bags, luggage, jackets, etc.

2. Satin Nylon: This irregular weave has fiber materials that pass over four fibers, then under one fiber. Satin Nylon is mostly used for jackets and has a smooth finish.

3. Taffeta Nylon- Taffeta Nylon has a dense weave where fibers are woven together. The smooth fabric allows for precise screen prints.

5.9   The problem

Printing on Nylon causes problems for every screen printer. Regular Dye will sit on top of the fabric versus bonding with the fabric when printed at a low temperature. Regular Dye bonds to fabric fibres when the temperature of the dye reaches 325 degrees Fahrenheit. This high temperature shrinks the nylon fabric. This is also because nylon material tends to be woven rather than knit. Although these problems exist, they can be solved with proper tools. When using the correct items and following the proper technique to screen print on Nylon. We can provide a crisp and clear print on Nylon fabric.

5.10   The solution

There are a few specialty items out there that can help solve the problem. Specialty ink, Jacket hold-down, retensionable screen, and sharp squeegee are items that will enable you to print on Nylon.

1. Fusion 180° Ink.- A plastisol-based ink is designed to print on Nylon. As I stated previously, regular plastisol ink bonds to fabric fibers when the temperature of the ink reaches 325 degrees Fahrenheit. This temperature melts nylon. With Fusion 180° ink the cure temperature is 240°F. The lower temperature allows the ink to bond to the material without melting the fabric. You will no longer need a catalyst, have to pre-heat the fabric prior to printing, wait 72 hours after the item was printed for the ink to be cured, or worry if the fabric will melt.

2. Hold-down. The jacket-hold-down will hold the nylon fabric in place. Nylon tends to have a smooth finish causing difficulties for the material to stay in place during the screen-printing process. This is especially true if a jacket contains a liner. Jacket Hold Down will prevent shifting and hold the fabric down securely.

3. A Sharp Squeegee.Squeegee must have a sharp, straight level edge. We recommend using a 70 or 80 single durometer squeegee.

5.11   Tips for screen printing on nylon

• Set Up. Off-contract distance should be 1/16" to 1/8". Flatten the nylon surface prior to printing, as the material tends to be quite wrinkly. This will guarantee that wrinkles will not become a part of your print. Due to the fact that Nylon shifts easily, you have only one chance to lay down the fabric on the pallet for a good print.

• Cure at a lower temperature. Nylon can shrink when the material is subjected to high temperatures. Our ink will allow you to cure the fabric at 240°F for 40 seconds through the dryer. You will not need a catalyst when using our ink nor will you have to wait 72 hours after curing. You will also be able to print nylon products that are waterproof. Use one stroke and do not cure with the flash.

• Cleaning. You will need to clean screens and squeegees immediately after printing. Our low temperature ink can dry out quickly so you will want to clean the ink out of all of your equipment as soon as your press run is complete.

• Test. Practice makes perfect. Make sure that you run a test print prior to completing the entire order.

5.12   Flat bed screen printing of textiles

Screen printing is by far the most popular technology in use today. Screen printing consists of three elements: the screen which is the image carrier; the squeegee; and ink. The screen printing process uses a porous mesh stretched tightly over a frame made of wood or metal. Proper tension is essential for accurate colour registration. The mesh is made of porous fabric or stainless steel. A stencil is produced on the screen either manually or photochemically. The stencil defines the image to be printed in other printing technologies this would be referred to as the image plate.

In flat bed screen printing, this process is an automated version of the older hand operated silk screen printing. For each color in the print design, a separate screen must be constructed or engraved.

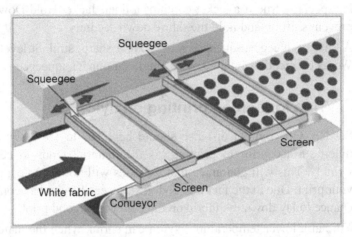

Figure 4.1 Industrial flad-bed screen printing
(From BBC, Bitesize, Design & Technology, Printing)

If the design has four colors, then four separate screens must be engraved. The modern flat-bed screen-printing machine consists of an in-feed device, a glue trough, a rotating continuous flat rubber blanket, flat-bed print table harnesses to lift and lower the flat screens, and a double-blade squeegee trough. The in-feed device allows for precise straight feeding of the textile fabric onto the rubber blanket. As the cloth is fed to the machine, it is lightly glued to the blanket to prevent any shifting of fabric or distortion during the printing process. The blanket carries the fabric under the screens, which are in the raised position. Once under the screens, the fabric stops, the screens are lowered, and an automatic squeegee trough moves across each screen, pushing print paste through the design or open areas of the screens. Remember, there is one screen for each colour in the pattern. The screens are raised, the blanket precisely moves the fabric to the next colour, and the process is repeated. Once each colour has been applied, the fabric is removed from the blanket and then processed through the required fixation process. The rubber blanket is continuously washed, dried, and rotated back to the fabric in-feed area. The flat-bed screen process is a semi-continuous, start-stop operation. Flat screen machines are used today mostly in printing terry towels.

Many factors such as composition, size and form, angle, pressure, and speed of the blade (squeegee) determine the quality of the impression made by the squeegee. At one time most blades were made from rubber which, however, is prone to wear aund edge nicks and has a tendency to warp and distort. While blades continue to be made from rubbers such as neoprene, most are now made from polyurethane which can produce as many as 25,000 impressions without significant degradation of the image.

From a productivity standpoint, the process is slow with production speeds in the range of 15-25 yards per minute. Additionally, the method has obvious design limits. The design repeat size is limited to the width and length dimensions of the flat screen. Also, no continuous patterns such as linear stripes are possible with this method. However, this method offers a number of advantages. Very wide machines can be constructed to accommodate fabrics such as sheets, blankets, bedspreads, carpets, or upholstery. Also, this technique allows for multiple passes or strokes of the squeegee so that large amounts of print paste can be applied to penetrate pile fabrics such as blankets or towels. Currently, approximately 15-18% of printed fabric production worldwide is done on flat-bed screen machines.

Flat-screen and rotary-screen printing are both characterized by the fact that the printing paste is transferred to the fabric through openings in specially designed screens. The openings on each screen correspond to a pattern and when the printing paste is forced through by means of a squeegee, the desired pattern is reproduced on the fabric. A separate screen is made for each color in the pattern.

Figure 4.2 Flat screen printing

A variety of different machines can be used for printing fabrics. The most commonly used are described below.

5.13   Flat screen printing machine

Flat-screen printing machines can be manual, semi-automatic or completely automatic. One type of machine, which is still commonly found in printing houses, can be described as follows. The fabric is first glued to a moving endless belt. A stationary screen at the front of the machine, is lowered onto the area that has to be printed and the printing paste is wiped with a squeegee. Afterwards the belt, with the fabric glued on it, is advanced to the pattern-repeat point and the screen is lowered again. The printed fabric moves forward step by step and passes through a dryer. The machine prints only one color at a time. When the first color is printed on the whole length of the fabric, the dried fabric is ready for the second cycle and so on until the pattern is completed.

5.14   Advantages of flat screen textile printing

In other fully mechanized machines all the colours are printed at the same time. A number of stationary screens (from 8 to 12, but some machines are equipped with up to 24 different screens) are placed along the printing machine. The screens are simultaneously lifted, while the textile, which is glued to a moving endless rubber belt, is advanced to the pattern-repeat point. Then the screens are lowered again and the paste is squeezed through the screens onto the fabric. The printed material moves forward one frame at each application and as it leaves the last frame it is finally dried and it is ready for fixation.

In both machines the continuous rubber belt, after pulling away the fabric, is moved downward in continuous mode over a guide roller and washed with water and rotating brushes to remove the printing paste residues and the glue, if necessary. After this, the belt is sent back to the gluing device. In some cases the glue is applied in liquid form by a squeegee, while in other machines the belts are pre-coated with thermoplastic glues. In this case the textile is heated and then it is squeezed by a roller or simply pressed against the rubber-coated belt, causing the glue to soften and instantly adhere.

After printing, the screens and the application system are washed out. It is common practice to squeeze the color from the screens back into the printing paste mixing containers before washing them.

5.15   Difference between flat screen & rotary screen printing

As most of the screen printing procedures are different; we can discuss about the difference between Flat & Rotary Screen Printing Process:

1. In Flat Screen Printing Technology the screen is flat and moves up and down. Beside, the Rotary screen is round in size and it rotates.

2. In Flat screen printing the Squeeze is used. In Rotary Screen Printing, just Roller is used.

3. Small width fabric (45-50″) is printed on Flat Screen Printing. On the other side, large width fabric is printed on Rotary screen printing.

4. In flat screen printing pattern, there are only 6 to 8 colour is available. But if you use the Rotary Screen printing, then you will be able to use 16 to 24 colours in a pattern.

5. Flat screen printing process gives a lower production rate. And the Rotary Screen printing is comparatively faster than Flat Screen printing and production rate is 3 times higher than that.

6. The initial investment cost for Flat screen printing machine is low, so that it's less expensive. But, the initial cost of investment in Rotary screen printing is too high; so, it's expensive.

Though, the Rotary screen printing machine is too much expensive; but in today's contemporary textile industry uses the expensive one because of it's productivity and flexibility.

5.16   Flat bed screen printing unit

The innovative rotatek flatbed screen printing unit is ideal both for printing fine lines, whole surfaces and relief printing up to a thickness of 300 microns. This gives the printer the option of highlighting even the tiniest detail or a volume and colour intensity unattainable with other printing technologies.

Its forward and reverse screen system with a fixed squeegee enables high speeds of up to 4,200 cycles/hour to be reached, the fastest on the market. The minimum print format is 200 mm, and the maximum 400 mm.

The main advantage of the screen printing system is its low cost compared with the rotary system.

The most notable innovation of the flatbed screen printing unit is that it can be integrated in-line with the full rotatek range of semi-rotary offset printing machines.

6
Discharge printing

6.1   Discharge style of printing

Discharge printing, also called Extract Printing, method of applying a design to dyed fabric by printing a colour-destroying agent, such as chlorine or hydrosulfite, to bleach out a white or light pattern on the darker coloured ground. In colour-discharge printing, a dye impervious to the bleaching agent is combined with it, producing a coloured design instead of white on the dyed ground.

The literal meaning of the word "discharge" is to eliminate or to remove. It means the style of printing which can produce a white or colored effect on a previously dyed fabric ground.

This discharging of color from previously dyed ground is carried out by a discharging agent, which is capable of destroying color by oxidation and reduction.

The discharging agents are – oxidizing agent (i.e. Potassium chlorate, Na-chlorate etc.) or reducing agent (i.e. Rongalite-c, Stannous chloride etc.)

6.2   Types of discharge printing style

2 types of discharging styles are available. White discharge and colored discharge.

6.2.1   White discharge

The fabric is dyed with certain class of dye and then printed (according required design) with a print paste containing "Reducing agent" (usually "Rongolite C") with no dyes. After printing, the fabric is steamed. The reducing agent on the printed portion destroys the ground color and thus produced a white on base color.

6.2.1   Color discharge

The fabric is dyed with certain class of dye and then printed (according to required design) with a print paste containing "Reducing agent" (usually

"Rongolite C") with dyestuff. After printing, the fabric is steamed. The reducing agent on the printed portion destroys the ground color and due to the presence of dye in print paste, the new color is replaced with the design area. Thus it produces a color-color combination.

Figure 6.1 White color discharge printed fabric

6.3   Discharge style of printing

This process is used to remove colour from a dyed fabric in a desired pattern. A paste containing a chemical with a bleaching effect is applied through a printing process. The fabric is then placed in a steam unit for a specified time. The steam has the effect of removing the colour from the fabric, leaving a lighter pattern in the printed areas. If desired, a non-dischargeable dye can be incorporated into the printing process to achieve a multicoloured design. In this method the fabric is pre-dyed to a solid shade by a traditional dyeing process and the colour is then destroyed locally, by chemicals incorporated in the print paste especially for that purpose. The result is a white patterned discharge on a coloured ground. In "white" discharge printing, the fabric is piece dyed, then printed with a paste containing a chemical that reduces the dye and hence removes the color where the white designs are desired. In "colored" discharge printing, a color is added to the discharge paste in order to replace the discharged color with another shade.

6.4   Dyed style of printing

This style consists of two steps: printing with a mordant and dyeing. After dyeing color only fixes in places where mordant was applied as a result a design is produced on the fabric.

6.5   Discharge style printing

Discharge means removal and discharging system means the process which can produce a white or colored effect on a previously dyed ground.

This discharging of color from previously dyed ground is carried out by a discharging agent which is actually a oxidizing and reducing agent capable of destroying color by oxidation and reduction.

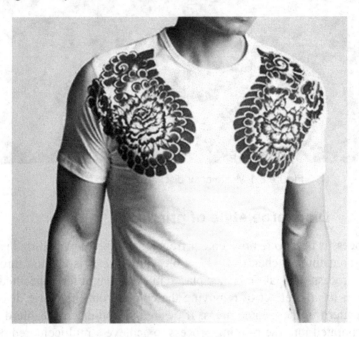

Figure 6.2 Discharge style of printing

Discharge styles have been important since the earliest days of textile printing..With any industrial process there must be sound technical and commercial reasons for its conception and continuation. In the case of discharge printing, the following considerations determine the usefulness of the process compared with other printing techniques.

1. Printed materials with large areas of ground colour can be produced, the depth, levelness and penetration of which would be difficult, if not impossible, to obtain by a direct printing process.

2. Delicate colours and intricate patterns can be reproduced on grounds of any depth, with a clarity and sharpness that have become the hallmarks of this style. Intricate white patterns lose their crispness if left as unprinted areas in a direct, blotch print, because the print

paste spreads unequally in different directions. In addition, a coloured motif fitted into a blotch print either leaves unprinted white margins or forms a third colour where fall-on occurs. In some cases such effects are acceptable, but they can be eliminated by using the discharge technique.

3. The extra processes required and the additional costs of discharge pastes mean that production costs are higher, but the aesthetically superior results give the product a higher value and enable profit margins to be maintained or even improved. The higher costs of discharge printing are often offset when applied to long-lasting designs used for scarves, ties, cravats and dressing gowns. As already indicated, in discharge styles the pattern is produced by the chemical destruction of the original dye in the printed areas. The discharging agents used can be oxidising or reducing agents, acids, alkalis and various salts. An early and, one might say, classical example is the discharge printing of cotton dyed with indigo, the characteristic colour of which can be destroyed either by oxidation or reduction.

Discharge printing is done as the following way.

Discharge Printing Discharge means to remove specific coloured area by another colour or reducing it by bleaching agent. There are two types of discharge printing, one is white discharge and another is colour discharge. In white discharge method, only dyed colour of fabric is removed whereas in color discharge, after removing the dyeing colour the required color is applied with discharging agent. Here, colour is destroyed by one or multiple colour. By this process pigment in the fabrics is removed chemically and replace it by another color. For discharge printing, fabric should be 100% cotton and the fabric should be dyed with dischargeable reactive dye .

Flow chart of pigment printing process :

Paste preparation (White paste/ Colour paste)

↓

Table preparation/ Machine preparation

↓

Fabric plaited on the table

↓

Pigment printing paste apply with the help of screen

↓

Curing at 190° C (belt speed 5 meter/minute)

↓

Delivery

6.6   Discharging agents

Clearly, the most important methods of discharging are based on reduction. This general method can be varied and adapted to give discharges with most classes of dye in use and on most types of fibre. Indeed, to many printers the terms 'reducing agent' and 'discharging agent' are synonymous. The most widely used reducing agents are the formaldehyde sulphoxylates. The stability of these compounds is such that only limited losses of sulphoxylate occur during printing and prior to steaming. The use of sodium formaldehyde sulphoxylate (CI Reducing Agent 2, sold as Formosul or Rongalite C) was established as long ago as 1905, when it was recognised that methods based on this reducing agent offered many advantages.

Advantages
- Large areas of ground of ground color are possible.
- Delicate colors and intricate patterns possible on deep ground color, excellent depth and clarity possible.
- Higher production cost but long lasting unique styles.

Disadvantages:
- It is an expensive process.
- Two stage application involved in dyeing or padding and discharge printing.
- Limited choice of ground and motif colours. Requires rigid process care that any default will lead to damages

6.7   Discharge printing

In discharge printing the ground color is applied by dyeing or padding with dis-chargeable dyes. It is a very cool technique that removes the dye from the fabric. Discharge printing is like to screen printing process but here instead of normal ink, discharge inks are used. Which remove the fabric's dye instead of putting a color on top of the fabric or shirt. Discharge printing works best on cotton fabrics because other fabrics are dyed in different processes and react less to discharge chemicals and therefore the dye does not break down.

6.8   Methods of discharge printing

Basic steps of discharge printing are printing, drying, steaming and washing. This technique is used on dyed fabrics (usually in dark shades). The fabric is dyed in the piece and then overprinted with a discharge paste (chemical) that destroys or decolorizes or changes the color in designed areas. A white discharge is when the original white is restored to the printed area. A color discharge is when a separate color is applied at the same time as the discharge paste. Sometimes the base color is removed and another color is printed in its place; but usually a white area is desirable to brighten the overall design.

Figure 6.3 Discharge printing on t-shirt

This printing method is generally used to obtain designs with tiny details, sharp and well defined edges on colored backgrounds, patterns with low coverage ratio on colored backgrounds, and to avoid pattern matching problems on endless design patterns with colored backgrounds. The results obtained with this printing method could be hardly reproducible with direct printing since it would be very difficult to obtain wide backgrounds, smooth and well penetrated, with sharp edges without seam defects.

A problem for this printing method is represented by the need to choose perfectly destroyable dyes for backgrounds, which cannot be affected by the discharging agent used as brightener. The selection restricts the number of applicable dyes and above all, for some color classes, very few dyes grant a good fastness to light and moisture, but excellent color effects. With this type of printing carried out on black or navy blue backgrounds it is also impossible to check if the various colors are correctly positioned; any mistake will be visible only after the steaming process and at that point it would be impossible to correct it. This problem could be limited by testing the printing result on a white cloth before beginning the printing process.

6.8   Discharge printing

Discharge printing is more difficult than normal printing. The results can often vary even after extensive testing.

The process allows for lighter colours to be printed on darker coloured backgrounds.

A bleaching agent is used to achieve this effect. The bleaching agent does not (usually) affect the fabric.

The base colour must be dischargeable whilst the printing colour must be non-dischargeable.

The colour range of non-dischargeable dyes is relatively small for cellulose fabrics compared to those available for protein fabrics.

It is possible to discharge print on most fabrics – so long as the correct choice of dye is used.

UK designers Bay & Brown use discharge a lot in their work.

The discharge agents used are:

Formosul – used for silk

Decrolin – used for wool and pigment printing

The discharge agents can be used on other fabrics such as cotton, but always test first!

6.9   Discharge printing

6.9.1   Basic recipe

Decrolin/ Formosul	20 gms
Glycerine	2 gms
Thickener	50 gms
Indalca water	28 gms
Total	100 gms

For wool and cotton use decrolin

For silk use formosul

This is a basic recipe that removes all or some of the dye from the fabric.

The colour left could vary from white to a pale version of the original colour.

Always test first and record what you have used and done.

Print

Dry

Steam – 20 minutes at 100 degrees

Wash – cold water for 20 minutes

Colour discharge printing
Recipe silk

Dye	2-5 gms
Formusol	20 gms
Glyzine	3 gms
Thickener	45 gms
Water	? gms
Total	100 gms

Use Acid dyes
Use either indalca or manutex thickeners
Print
Dry
Steam – 100 degrees 30 minutes
Wash – use plenty of cold running water.

Colour discharge printing
Recipe wool fabrics

Dye	2-5 gms
Decroline	20 gms
Glyezine	3 gms
Thickener	55 gms
Water	? gms
Total	100 gms

Use Acid dyes
Use Indalca Thickener
Mix the dyes, chemicals with water thoroughly, then add the thickener.
Print
Dry
Steam – 100 degrees 30 minutes
Wash – use plenty of cold running water.

Pigment colour discharge printing
Recipe cotton

Pigment colour	6 gms
Decroline	8 gms
Discharge base	86 gms
Total	100 gms

This recipe works well on non-colour fast dyed cotton, viscose.
This recipe can work on silk and wool.
Once mixed, wait 15 minutes before printing.
Print

Dry
Steam – 100 degrees in steamer
Bake – 165 degrees in baker or transfer press

Color discharge printing

It is a process where printing of lighter colors is done onto fabrics having darker background. It works best on cotton fabrics because other fabrics are dyed in different processes & react less to discharge chemicals & therefore the dye does not break down.

Figure 6.4 Color discharge printing

Discharge printing techniques were particularly fashionable in the 19th century. It has been important since the earliest day.

Discharge printing, also called Extract Printing, method of applying a design to dyed fabric by printing a color-destroying agent, such as chlorine or hydrosulfite, to bleach out a white or light pattern on the darker colored ground.

6.10   Steps involved in color discharge printing

- Preparation of fabric to be printed.
- Dyeing of fabric with dischargeable dye.
- Preparation of discharge paste.
- Printing the discharge paste onto the dyed fabric.
- Drying.
- Steaming under atmospheric condition and under pressure.
- Washing and soaping as required.
- Drying.
- Finishing.
- Packing, billing and shipping.

6.11   Print pastes for discharge printing

A typical print paste for discharge printing contains dye (for colored discharge), discharging agent, thickener, other chemicals and auxiliaries and water. Each component requires detailed consideration.

6.12   Dye selection

The correct choice of dye is of fundamental importance in successful discharge printing, with respect to both the dischargeable ground and where required, the illuminating, discharge- resistant dye.

Dyes which are suitable for the dischargeable ground usually contain azo groups that can be split by reduction. Even so, there are great differences in discharge ability between individual dyes.

Whereas, the great majority of discharge-resistant dyes are not azo dyes but are of the anthraquinonoid, phthalocyanine and triphenylmethane type. The choice depends upon the color required, the reducing agent being used and the substrate.

6.13   Classification of dye according to discharge
ability scale

The choice of dyes is facilitated by the dye manufacturers, who usually classify their products on a discharge ability scale ranging from 1 to 5. A dye which is classified as 5 or 4-5 on this scale would be suitable for a white discharge. For a colored discharge, a dye classified as 4 would be acceptable or even 3-4 with very deep illuminating colors. Those dyes which have a discharge ability of only 1 are virtually undischargeable and therefore, are suitable as the illuminating colors in colored discharge styles.

Only a trial under actual working conditions can provide full and final information on the suitability of a dye for discharge printing.

6.14   Discharging agents

The most important method of discharging is based on reduction. This general method can be varied and adapted to give discharges with most classes of dye in use and on most types of fiber. Indeed, to many printers the terms 'reducing agent' and 'discharging agent' are synonymous.

The most widely used reducing agents are formaldehyde suphoxylates. The stability of these compounds is such that only limited losses of sulphoxylates occur during printing and prior to steaming.

Another reducing agent, which has been used since the earliest times is the Tin (II) Chloride. It is a readily soluble compound which reacts with an azo dye. The importance of Tin (II) Chloride diminished considerably on the introduction of the sulphoxylates, but it has now regained some significance in discharge printing of synthetic fibers.

The choice of reducing agent is determined largely by the fiber to be printed and, to some extent, by the dyes used. The soluble sulphoxylates can give haloing problems on synthetic fibers, caused by capillary movement of solution along the yarns. This problem can be overcome by using the insoluble formaldehyde sulphoxylates.

Generally, sulphoxylates are stronger reducing agents than Tin (II) chloride, and can be used to discharge a greater range of dyes. On the other hand, since very few dyes are absolutely resistant to reducing agents, tin (ii) chloride is preferred with illuminating dyes.

Amount of reducing agent depends on:

The actual amount of reducing agent required for optimum discharge will depend upon:

- The dyes to be discharged.
- The depth of the ground.
- The fabric being printed.

6.15   Excess and Insufficient reducing agent causes

The use of too much reducing agent results in flushing or haloing during steaming; as well as being wasteful and uneconomic. Flushing of a white discharge results in blurred edges and a loss of fine detail, whilst in colored discharges it is usually seen as a white halo around the printed areas.

Whereas, use of insufficient reducing agent will, of course, give an incomplete discharge.

6.16   Thickeners

The correct choice of thickener is of great importance as they act as vehicle for carrying dye onto the cloth. The thickener must have good stability to reducing agent used, otherwise coagulation may occur.

6.17   Which thickener should be selected?

Following points should be considered while selecting a thickener.

- Non-ionic thickeners are necessary. Anionic thickeners should be avoided such as carboxymethylated types.
- Since sharp and intricate patterns are characteristic of a discharge style, it is essential to minimize flushing and bleeding. It is therefore, necessary to use low-viscosity thickeners and a high solid content.

6.18   Other chemicals & auxiliaries

6.18.1   Catalyst

Anthraquinone is often used to improve the discharge effect of a reducing agent, and is therefore used on fabrics dyed with the azo dyes which are more difficult to discharge. During steaming it is reduced to hydroanthraquinone which, in turn, reduces the dye and is itself reconverted to Anthraquinone. This cycle of reaction continues until reduction of dye is complete. The Anthraquinone might be therefore considered as a catalyst. To prevent subsequent discoloration, all Anthraquinone must be removed in the washing process that follows.

6.18.2   Penetrating agents

It is often necessary to employ penetrating agents during steaming, to ensure that discharge paste thoroughly penetrates the fabric. Additive of this type include glycerol, ethylene glycols and thiodiglycols.

6.18.3   Carriers

Carriers and Fixation accelerators are often added when printing illuminated discharges on synthetic fibers, in some cases, they improve a white discharge on such substrates.

6.18.4   Wetting agents

These are also necessary when printing on fabric of low absorbency which may be coated with dried film of thickener from the preliminary dyeing operation, as in the 'discharge-resist' process.

6.19   Coloured discharge with vat dyes

6.19.1   What are vat dyes?

In textile printing, vat dyes plays a very important part, vat dyes are used for both for the ground shades and discharge colors. Vat dyes in pigment from are insoluble in water and do not have affinity for cellulosic fibers so, we have to dissolve the dye and the process of dissolving the dye is called Vatting.

The coloured ground is applied using selected azoic, direct and reactive dyes while the illuminating colours are selected vat dyes.

- Preparation of the dyed ground fabric in carried out by padding with "resist salt" and dried.
- After printing and drying the print are steamed for 5-8 min at 102-104 °C in air free-steamer.
- Washing and after treatment are carried immediately.
- The dyed fabric is then processed to achieve the discharge.
- First, the colour chromophore is reduced and eliminated.
- Then the dye linkage to the fiber is broken and these by-products of discharge are removed in washing.
- The discharging agent simultaneously reduces the vat dye, enables it to colour the fibre then converted into the original insoluble form. The first boxes of the washing range are used for oxidation of the vat dye.
- Hot soaping is then done with addition of suitable detergent and soda ash, If neccessary.

6.19.2   Why vat dyes are used with reactive, azoic and direct dyes?

The improved cellulosic fibres can be prepared by impregnating a monomers containing quarternary amine group into cellulosic fibres and polymerizing the monomers in the fibres. This treatment can be advantageously utilized in discharge printing of a fabric from a blend of hydrophobic synthetic/cellulosic fibres.

In the colour discharge printing of fibres,a vat dye or pigment is generally used as the effect of colour of printing to be added to the discharge printing paste.

6.20   Problems in using vat dyes

When we use vat dye as effect colour of printing and sodium formaldehyde sulphoxylate, which is usually employed as reducing agent is unstable and thus its reducing capacity is decrease after the paste is applied to the material to be printed, thereby making printing technically difficult and also making the acquisition of clear colour print difficult.

When pigment is employed the resulting prints are liable to have a hard hand and paint like colour shade which degrade the quality of product.

6.21   Advantages of coloured discharge printing

Some advantages of coloured discharge printing:
- Large areas of ground colour are possible.
- Delicate colours and intricate patterns possible on deep ground color, excellent depth and clarity possible.
- Higher production cost but long lasting unique styles.

6.22   Disadvantages of coloured discharge printing

Color discharge printing also has some disadvantages:
- It is an expensive process.
- Two stage application involved in dyeing or padding and discharge printing.
- Limited choice of ground and motif colours. Requires rigid process care that any default will lead to damage.

6.23   Conclusion

In recent years, modern techniques have made the use of direct printing practicable for many more designs and reduced the necessity of using these styles but they will always be significance because the effect obtained are often different and aesthetically superior.in discharge printing the fabric must be dyed with dyes that can be destroyed by selected discharging agents. The discharging paste is printed on dyed fabric and usually during subsequent steaming the dye in the pattern area is discharge.

6.24   New method of discharge printing on cotton fabrics using enzymes

Discharge printing is a method where the pattern is produced by the chemical destruction of the original dye in the printed areas. The discharging agents used can be oxidizing or reducing agents, acids, alkalis and various salts. But, most important methods of discharging are based sulphoxylates formaldehyde

& Thiourea dioxide. Discharge style of printing is done using sodium hydrosulphite, stannous chloride and zinc sulphox-ylate formaldehyde as discharging agent on reactive dyed cotton fabric. With sodium hydrosulphite as a discharging agent, excellent colour fastness to rubbing in both dry and wet conditions is achieved. Colour fastness to light and washing is good and does not vary with diff erent discharging agents used in discharge style of printing. CIE Whiteness Index reduces after application with diff erent discharging agents. However, whiteness index of reactive dyed cotton fabric discharged with sodium hydrosulphite shows superior value in comparison to stannous chloride and zinc sulphoxylate formaldehyde at the white discharged areas

Recently, the environmental and industrial safety conditions increased the potential for use of enzymes in textileprocessing to ensure eco-friendly production. In Discharge printing, Sulphoxylate Formaldehyde ($NaHS0_2.CH2o.2H20$) is one of the powerful discharging agent used commercially, however, it is considerably toxic and evolves formaldehyde known as a human carcinogenic associated with nasal sinus cancer and nasopharyngeal cancer.

In this article we have discussed about replacement of this hazardous chemical with eco- friendly enzymes in textile discharge printing. Enzymatic discharging printing carried out with Phenol oxidizing enzyme such as Peroxidase with hydrogen peroxide by selectively discharged reactive dyes from the cotton fabric at selected areas creating a printed surface.

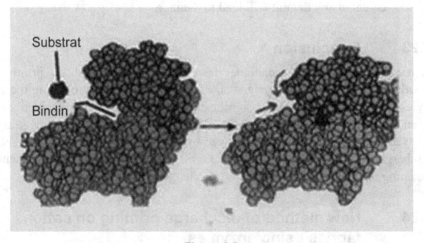

Figure 6.5

Textile discharging printing is the most versatile and important of the methods used for introducing design to textile fabrics. Considered analytically it is a process of bringing together a design idea, one or more colorants, and a textile fabric, using a technique for applying the colorants with some precision.

Biotechnology has dramatically increased the scope for application of enzyme systems in all areas of textile processing. Enzymes can be tailored to implement specific reactions, such as decomposition, oxidation and synthesis, for a variety of purposes. There is a growing recognition that enzymes can be used in many remediation processes to target specific purpose in textile industry. In this direction, recent biotechnological advances have allowed the production of cheaper and more readily available enzymes through better isolation and purification procedures.

6.25   Enzyme

The enzymes are used in textile industry because it operates under mild conditions of temperature and pH, it replaces non-selective harsh chemicals. A Bleaching enzyme such as Peroxidases together with hydrogen peroxide is capable of oxidizing organic compounds containing phenolic groups.

Bacterial system present (Peroxidase) in activated sludge, they catalyses the oxidative cleavage of Azo dyes.

E + S $\longrightarrow$	ES $\longrightarrow$	P
(Inital state) $\longleftarrow$	(Intermediate state)	(Final state)

Enzymes are specific that is, they control only one particular chemical change or type of change.

Textile discharge printing is the most versatile and important of the methods used for introducing design to textile fabrics. In discharge styles, the pattern is produced by the chemical destruction of the original dye in the printed areas. The discharging agents used can be oxidising or reducing agents, acids, alkalis and various salts. However, the most important methods of discharging are based formaldehyde sulphoxylates and thiourea dioxide. Recently, environmental and industrial safety concerns have increased the potential for the use of enzymes in textile processing to ensure eco-friendly production. Formaldehyde sulphoxylate (NaHSO2.CH2O.2H2O) is one of the most powerful discharging agents; however, it is quite toxic and produces formaldehyde, a known human carcinogen associated with nasal sinus cancer and nasopharyngeal cancer. In this work, a hazardous chemical has been replaced with eco-friendly horseradish peroxidase enzyme in textile discharge printing. Enzymatic discharge printing was carried out with a phenol oxidising enzyme system such that the reactive dye was selectively discharged from the cotton fabric in selected areas, creating a printed surface. The effects of enzyme concentration, pH of the printing paste, treatment time and the temperature of enzymatic treatment were studied. The optimum conditions for enzymatic discharge printing were found to be pH 8.5 at 70°C with a dye concentration of 80 g/L and 60 min as the treatment time.

In the recent times, the environmental and industrial safety conditions have increased the potential for use of enzymes in textile processing to ensure eco-friendly production. Formaldehyde sulphoxylate (NaHS02.CH20.2H20) is one of the powerful discharging agents; however, it is considerably toxic and evolves formaldehyde known as a human carcinogenic associated with nasal sinus cancer and nasopharyngeal cancer.

This hazardous chemical has been replaced with an eco-friendly horseradish peroxidase enzyme in textile discharge printing. Enzymatic discharging printing carried out with phenol oxidizing enzyme systems such that reactive dye is selectively discharged from the cotton fabric at selected areas creating a printed surface. The effects of enzymes concentration, pH of the printing paste, treatment time and temperature of enzymatic treatment have been studied. The optimum conditions for enzymatic discharge printing are found to be pH 8, 80°C temperature and 30 minutes time of treatment.

Textile discharge printing is the most versatile and important of the methods used for introducing design to textile fabrics. Considered analytically it is a process of bringing together a design idea, one or more colourants, and a textile fabric, using a technique for applying the colourants with some precision.

Biotechnology has dramatically increased the scope for application of enzyme systems in all areas of textile processing. Enzymes can be tailored to implement specific reactions, such as decomposition, oxidation and synthesis, for a variety of purposes. There is a growing recognition that enzymes can be used in many remediation processes to target specific purpose in textile industry. In this direction, recent biotechnological advances have allowed the production of cheaper and more readily available enzymes through better isolation and purification procedures.

Discharge styles show the pattern is produced by the chemical destruction of the original dye in the printed areas. The discharging agents used can be oxidizing or reducing agents, acids, alkalis and various salts. But, most important methods of discharging are based on formaldehyde sulphoxylates & Thiourea dioxide.

Horseradish peroxidase is a protein with a molecular weight of about 40,000, which contains a single protoporphyrin IX hemegroup. This enzyme catalyses the oxidation of a variety of substrates by hydrogen peroxide. This present work is aimed at using horseradish peroxidase enzyme instead of toxic reducing agent to create discharge style on cotton fabric dyed with vinyl sulphone reactive dyes.

6.26   Structure feature of horseradish peroxidase

The enzyme horseradish peroxidase (HRP), found in horseradish, is used extensively in biochemistry applications primarily for its ability to amplify a weak signal and increase detect ability of a target molecule.

The first one – electron reduction step requires participation of a reducing substrate and leads to the generation of compound II, an Fe (IV) oxoferryl species that is one oxidizing equivalent above resting state. Both compound I and compound II are powerful oxidants, with redox potentials estimated to be a close to +1V. The second one -electron reduction step returns compound II to the resting state of the enzyme.

6.27   Trials taken for bio –discharge printing

The fabric dyed with reactive dye using procedure recommended by dye manufacturer. Exhaust dyeing was carried out at liquor ratio 1:30. Dyeing of fabric was carried out at 60oC for 60 minutes. Fixation was conducted for 20 minutes using 6 to 8 gpl of Na2CO3 and 0.01 to 0.5 gpl of caustic lye. Process conditions for dyeing are given in Table 3.

6.28   Bio-discharge printing

The cotton samples were printed with a printing paste using hand screen printing technique as per below recipe (White Discharge Printing):

Horseradish Peroxidase: 30, 50, 70, 80, 90 g/Kg

Sodium Alginate: 20 g

Hydrogen peroxide: 20 ml

The printed cotton samples were allowed to dry at ambient condition; then it was left in an oven for different intervals of time and at different temperatures. Finally washing was carried out.

6.29   Washing

The printed fabrics were rinsed with cold water followed by washing in presence of sodium perborate at 60° temperature for 30 minutes. After the washing, the incubated fabrics were again washed with ECE detergent (4g/L) at 60° temperature for 30 minutes. Finally samples kept in air drying room.

Horseradish Peroxidase Enzyme has been used instead of toxic reducing agents, viz, Formaldehyde sulphoxylate which can release formaldehyde known as a human carcinogenic under a variety of conditions. The effect of enzyme concentration, pH, treatment time & temperature on white discharge performance has been studied.

Abrasion resistance result is also low in conventional printed fabric. Fastness & absorbency results of fabric are same in both printed methods.

Bio-technology & enzyme application is inevitable tool in modern industry where environmental aspect plays critical role to sustain in the competitive market. Innovative method of using Horseradish Peroxidase & H2O2 formulation in discharge printing of textiles carried out successfully. Formaldehyde liberation can be fully avoided in this kind of bio-discharge printing.

The optimum conditions for using Peroxidase formulation are found to be pH 8, 80°C temperature and 30 minutes incubation time. By using peroxidase enzyme discharging printing method, the following advantages have observed.

* Elimination of formaldehyde
* Energy saving
* Reduction of strength loss
* Environmental-friendly.

7

Block printing

Block Printing is one of the oldest types of printmaking, and has been around for thousands of years. There is evidence that it existed as early as the fifth century BC, with actual fragments found from as early as the fifteenth century. It has been done around the world, with roots in India, China and Japan. The art of Indian block print textiles is a labor-intensive, painstaking process that has survived from ancient times to the present because of the beauty of the handmade products. Scraps of cloth found in the ruins of Mohenjo Daro, an ancient city of the Indus Valley Civilization, provide evidence that block printing was practiced in India as long ago as 3000 BCE. The art flourished in the 12th century under the patronage of the rajas. The 17th century saw a revitalization of the art. And still, here in the 21st century, block printing of fabric by hand is an art practiced by Indian artisans for the enjoyment of owners of those fabrics throughout the world. Block printing is an ancient textile tradition that originated in the Rajasthan Desert region of India centuries ago. It involves stamping designs onto fabric by hand using intricately carved wooden blocks. Block printing is a labor-intensive, painstaking process which requires time, teamwork and a tremendous level of skill. For example, it can take five carvers up to three days to create a pattern in a printing block. And the printers may use as many as thirty blocks to complete one design!

Despite competition from faster and cheaper methods of textile design, block printing has resisted industrialization throughout the years and is still done without any mechanization. It has endured in popularity because of its natural feel and its ability to achieve a three-dimensional, artistic aura which is difficult to replicate in machine-made and mass-produced items.

Block-printed textiles are popping up in everything from high-end fashion and home decor to DIY projects on Pinterest. However, block printing is not a newfound trend; it is an age-old art with deep roots in ancient civilizations ranging from China and Egypt to Assyria. For centuries, block printing has been used to embellish religious cloths, billowing skirts, saris, and robes for royalty. Block-printed fabrics showcase the perfect imperfections of the human hand– the sensibilities and skills of real craftsmanship. To put it simply, every bolt of fabric, scarf, or sarong made using this technique is special and distinct.

The process of block printing takes time, teamwork, and a high degree of skill–required for both the placement of motifs and the application of pressure. Depending on the amount of pressure that's applied to the block stamp, each piece of fabric showcases varying levels of color depth and dye absorption–taking on the style of the artisan who created it. This unique and time-intensive process means no two pieces of fabric are the same.

Since there is such a long history of block printing, there are many different techniques, but it is essentially using a carved material covered in ink to transfer an image on to fabric. Block printing can be done with wood, linoleum, rubber, or many other materials.

There are various types of process followed at different places

7.1   Process A

- **Step 1:** A design is drawn or transferred onto a wooden block called a bunta.
- **Step 2:** The wood block is then hand carved, leaving the most delicate work until last to avoid the risk of damaging it. Trained craftsmen make the blocks out of seasoned teak wood. Various point marks are carved into the block which printers use as placement indicators. When finished, the block presents the appearance of a flat relief carving. Each block is soaked in oil for one to two weeks to soften the wood before use.
- **Step 3:** The printer applies color to the block and presses it firmly on the cloth. He or she ensures a good impression by striking it on the back with a wooden mallet. The second impression is made in the same way, with the printer taking care to see that it fits exactly to the first. Each succeeding impression is made in the same manner until the length of cloth is fully printed. Printing is done from left to right. If the pattern contains several colors, the cloth is usually printed first with one color, dried, then printed with the second color. (The same operations are then repeated until all of the colors are printed.)

7.1.1   The process

The three main tools of block printing are the wooden blocks, the fabric and the dye:
- Raw fabric is first washed to remove starch and then bleached in a gentle solution.

- Once bleached, it is hung up on tall bamboo frames to dry in the hot desert sun.
- The blocks are made from teak wood by trained craftsmen. Designs are traced onto the surface and then carved with a hammer and chisel.
- Each block is made with a wooden handle and several small holes to release air and excess dye. It is soaked in oil for 10-15 days before use to soften the grains of the timber.
- Colors are mixed in a separate room from the printing. They are then kept on a tray which the printer drags along as he works.
- In preparation for printing, the cotton fabric is stretched tightly over the printing tables and fastened with pins to avoid wrinkles and movement.
- When printing begins, artisans first print the outline of the design. This is usually done by the most experienced printer, as the outline leads the whole process and must be very precise.
- The rest of the team then fills in the colors with the various "fill" blocks.
- Once printed, the swaths of fabric are hung to dry in the sun before a final wash.

The fact that block printing is still in vogue after centuries of use is a true testament to its distinct, unique beauty. Perfection has never been the goal—every time the fabric is hand stamped, it is unique. Perhaps it is this authentic, rustic feeling that consumers crave, a lean towards the past and a nod to the traditions of our ancestors.

The process of block printing takes time, team work and, especially, skill. The three main tools of a block printed fabric are the wooden blocks, the fabric and the dye. It can take five carvers up to three days to create an intricate design in a block of teak for use as a printing block. The printers may use up to 30 blocks to complete a design. Separate blocks are required for each of the colors used in a design and it is not unusual to have four or five colors in a professional design. It can take twenty people, each doing a separate task, up to eight hours to prepare a single block printed garment. With all this, the results can only be unique.

The process of block printing begins with the wooden blocks. Wood carvers cut designs into blocks of different shapes and sizes. The top of the block has a handle for the printers to grasp. Each block has two or three cylindrical holes through it to permit the passage of air and to allow excess dye to squeeze out. There are also various points carved into to block which the printers use as placement indicators as they pick the block up and move it to the next patch of fabric. Each block is soaked in oil for one and one half to two weeks to soften the wood.

The main tools of the printer are wooden blocks which are available in different shapes and size as per the design and requirement. The block makers mainly make two types of block

1. Wooden Block
2. Metal Block

7.1.2   Wooden block

These blocks are usually made on teak or *seesham* wood. Artisans make sure that the wood is seasoned and then carve the motifs on it. The design are first printed on paper and stuck on the block of wood. Artisan, then start carving the wood with steel chisels, of different widths and cutting surface. The motifs are carved on the base while the top has a handle. These handles are either carved out from the same wood or by a low cost wood attached to the surface with the help of nails. Each block has two or more cylindrical holes drilled into the block for free air passage and also to allow release of excess color. Blocks made are of rectangular, square, oval, semi-circular, circular etc. shape. Once the block is made it is soaked in oil for 10-15 days to soften the grains. The life of these block are approximately 600-800 meter of printing. The outline blocks are called as *rekh* and the filler blocks are called as *datta*.

7.1.3   Metal block

For making intricate patterns and getting high level of clarity in prints metal blocks are used. These blocks are made by engraving thin sheets of evenly cut metal strips into the wooden blocks. The metal strips are beaten to make them thin and then strips are cut of even length. The design is drawn on the wooden block and the metal strips are pressed onto the design by gentle hammering. Filling of the designs is done from center to outside. Once made the blocks are checked to see the strips are of the same height from the wooden base. Metal blocks are costly, time consuming but long lasting.

The process of printing can be divided into following major parts:-

- Washing
- Marking
- Printing
- Drying
- Washing

Washing: The fabric brought from the market contains starch, dust etc. Therefore it goes through a preprinting treatment in order to get good results while printing. The fabric is therefore dipped in a solution of water and bleach for 1-2 days. It is then boiled and washed with water. Finally the cloth is stretched and spread on ground and left for drying. This process known as Hari Sarana takes 3-5 days depending on the weather conditions.

Marking: The cloth to be printed is spread on the printing table and fixed with the help of pins. With help of scale and chalk areas to be printed are marked and proper gaps are left for cutting and stitching.

Printing: These are specially made tables measuring approximately 3 feet high, 3 feet wide and 9 feet long. The surfaces of the tables are covered with several layers of cloth, jute and canvas which help in better printing. These tables allow 3-4 printers standing and working simultaneously. Traditionally these tables were of low height approximately 1 feet high, 2 feet wide and 5 feet long. The printer used to sit on ground and print.

Color Plates: Colors used for printing are kept in a wooded tray called as *saaj*. Once the color is poured into the tray wire mesh is placed inside. This mesh is covered with a piece of felt. Felt soaks the color and finally a fine cotton/*malmal* cloth is placed on the felt. This preparation helps in proper application of color on the block.

Tray Trolley: These are wooden trolleys' accommodated with two shelves and wheels in the base for the easy movement. On the upper shelve the color tray is kept while in the lower rack required blocks. The height of the trolley is 3 feet suitable for working on the printing table. These are locally called as patiya.

Scale: For marking the areas to be printed scales are used.

Chalk: For marking chalks are used. These chalks have sharp edges which gives fine line.

Brushes: Metal or nylon brushes are used to clean the wooden and metal blocks after use. This helps in maintaining the life of the block.

Tambadi (Copper vessels): Traditionally copper vessels are used for dyeing and washing of cloths.

Mogari (wooden roller): This a cylindrical wooden roller on which the cloth is kept and beaten.

Kotan (Wooden mallet): This is used to beat the cloth over mogari, in order to remove the starch.

Printer dips the block in the color and stamps the design on the cloth. The blocks are then pressed hard with the fist on the back of the handle so that registration/imprint of the color is even. Printing starts from left to right. Number of colors used in the design defines the number of blocks to be used. Generally one printer handles one color and application is done simultaneously. In the case of sarees the pallu is printed first and then the border. First the outline color is applied and then the filling colors. Specific point in the block guides the printer for the repeat impression. The process of printing is called as Chapaai.

7.1.4   Raw materials

The next step in the process is the arrangement of the fabric. Workers stretch 24 layers of jute taut over a long rectangular table. The jute serves as a pad to provide resiliency to the printing surface. The workers secure the fabric to the jute pad with pins, keeping it tightly in place.

The process of block printing is widespread due to its intricate process, motifs and vibrant colors. The main raw material is the color used in the printing. Traditionally the artisans used natural colors but today it has been replaced by chemical and artificial colors.

When the printers are ready to do the printing, they select from three approaches. In the first method, called discharge printing, the printer dyes the fabric first. Then the printer chemically removes the dye from the portions of the cloth which will take the design. The bleached sections are treated, then dye is reapplied to create the block print design. In the second method, also known as direct block printing, the cloth is bleached, then dyed whole. The background color remains as the printer proceeds to print designs onto the dyed fabric using the wooden blocks. Finally, in resist printing, the block is used, not to apply dye, but rather to apply an impermeable resist, which can be made of clay, resin or wax. When the cloth is dyed, the portions covered by the resist do not take the dye. The resist is removed and the design has been created in reverse.

Printing is done from left to right. When the printer is using the discharge or direct block print methods, the printer dips the block into the dye then presses it onto the fabric. The printer slams the back of the block hard with the fist to create a clear impression. Then the printer moves the block to the next portion of fabric to be dyed, using points on the block to serve as a guide for the placement of the block.

As they work, the printers pull a wooden cart bearing their blocks along with them. The wooden blocks can be interchanged from one piece of fabric to another, creating different patterns. Custom designs and different colors can be used from one fabric to another, creating still more individual work.

Each color of a design is done by a different printer, coming behind the one before and repeating the process. The process requires teamwork, as each subsequent printer must place the block accurately to create a beautiful, whole pattern.

Once the pattern is finished on the whole length of fabric, the piece is treated to fix the dyes. First, the fabric is dried in the sun. Once dried, the fabric is rolled in newspapers and steamed in special boilers. After steaming, the fabric is washed, dried in the sun again, and ironed. Each of these steps contributes to fixing the pigment and making the colors rich and vibrant.

And it seems to work. This process of block printing has been used for centuries and is still in vogue.

The main raw materials are Colors/ Dyes.

Different types of dyes are used for silk and cotton.

* Vegetable/ Natural dyes
* Discharge Dyes
* Pigment dyes
* Reactive dyes
* Rapid fast Colors

7.1.5   Vegetable/ Natural dyes

Since ages artisans are using vegetable dyes for printing and dying. As they are extracted from the nature, vegetables, fruits etc., they are beneficial for the environment along with having traditional importance. There are few major colors produced naturally which are known internationally too.

7.6.5.1   Indigo blue

Extracted from the indigo plant found throughout India.

7.6.5.2   Red

This is obtained by mixing alizarin with alum. The color ranges from pink to deep red.

7.6.5.3   Black

This is acidic solution of iron which is obtained by processing rusted nails, horse shoes, iron scrap etc. with jiggery and salt. The mixture is buried under the ground and allowed to rot for about 10-15 days. It is then taken out and the color is prepared.

7.6.5.4   Green

The skin of pomegranate is processed by boiling.

Apart from this artisans also use bark of mango tree, vinegar, slaked lime etc.

7.1.6   Discharge dyes

These dyes are used to print on dark background. The printing colours have chemical which react with the dark ground fabric and bleaches out the dark colour from that particular place and prints the desired colour. In this process a range of white and other light colours can be printed on a dark background. Some Dyes Company are manufacturing and selling Dischargable Reactive Dyes.

7.2   Hand block printing in India

7.2.1   Block printing

Figure 7.1 Block prints on tussar silk saree

Remember, when you were 8 years old, you took out a ladyfinger cut it in two halves dipped it in water color and painted the canvas with the colors of your choice. That is how you can infer hand block prints technique to be like.

The intricate designs that you see printed on the fabric are not designed using a brush. They are patterns printed using wooden blocks. The fine and adorable prints in the shape of flowers, leaves, animals and abstract designs, give a simple cloth an interesting face.

7.2.2   Origin and history

The art of creating block print that is equal and aligned was a requirement felt quite early in the history of the Indian fashion industry. The hand block printing technique was created somewhere as early as 3000 BC. History acquaints us with the art as a tradition followed by the civilization of Mesopotamia. The engraved patterns on the coins used at that time are archaeological examples of the existence of block printing at that time.

The images and block prints that are widely used in East India as a style quotient originated in China. Before this technique came to India, it was a well-established art in the continent of China. The art flourished in the era of maharajas, which happened somewhere around 12th century. However, the practice was conceived long before it got imported to India. Since its inception, the art has been an integral part of the apparel design industry.

7.2.3   Present day scenario

Figure 7.2 Block prints

It has been ages since the idea of block printing became a reality. Today, the art of block printing has taken a different route. Block printing has been an elaborate technique since the very beginning. However, it did not have many facets back in time. Hand block printing is one of the most admired arts in the region of Rajasthan. Because of its tremendous demand, the fabrics that have been block printed are exported to other regions in bulk.

7.2.4   Sources of inspiration

Block printed designs are enamoring. The imprints when impressed on the cloth add an amusing allure to the cloth. Be it bed linen, fabric for apparel or the curtains for your wall, hand block printing is the choicest style for innumerable years in the Indian handloom industry. Unlike the Phoolkari or most other design trends, hand block printing takes its cue from the age old form of arts like Mehendi Designs, and from abstract designs. A wide variety in design and the creative ability of a designer enables a range of patterns varying from simple waves and circles to more elaborate designs. The smaller patterns were at a time known as 'Chintz'. 'Buti', Buta, 'Jhal', 'Bel' formed the components of each block print.

7.2.5   Faces behind the fabric

Unlike most other forms of design, hand block printing is profoundly embossed in the history and culture of almost all Indian states. You have the block printing of Punjab, Block printing of Rajasthan, Block Printing of Gujarat, and Block printing of Andhra, to name a few.

Figure 7.3 Hand block print (Image: BP Blog)

7.2.6 Varieties

The wooden blocks are left to the imagination of the carver who creates intricate patterns of the block. These could be simple geometric formations or more detailed impressions. Depending on the pattern that the artist wants to create on the fabric, the block prints are used on the cloth like **block print sarees** and several other clothes.

The workers of these states unite to form an industry that is extremely important to Indian and global fashion. While the technique of hand block printing has remained uniform throughout the years and across the states, the variety results from the intelligent and creative usage of blocks to create interesting patterns.

7.2.7 Innovations

From table napkins to canopied light-bulbs, everything today can see a bit of block print on it. Hand block printing, as a technique, has touched every industry and perforated in its roots to become an indispensable trend.

Figure 7.4 Printed cotton front slit kurta in teal blue

The innovations that have been introduced in the industry are not just an about the varied facets of design. They are also about the shift in technology that the block printing industry has witnessed. When screen printing invaded the industry, it ate up the entire hand block printing industry. Now that people have started understanding the innate differences between block and screen print, both industries flourish in co-existence.

Today, block printing beats the demand for screen prints because of the eco-friendliness that the former has introduced in its practice. The colours and dyes that the industry utilizes are free from any synthetic substance making is an obvious choice of the consumer.

Figure 7.5 Blocks for printing

7.2.8   Global appeal

Hand printing fits miraculously well into the India as well as Western culture. From salwar suits to trendy tops, block prints can be found adorning the neck-lines and hemline of a range of apparels. The global wear-ability of this fabric can be owed to its chic and sophisticated appeal. Moreover, the fact that the maintenance of block print is easy draws consumers from across the globe.

7.2.9   The future

The future of block printing looks quite impressive from the point where the industry stands today. Hand block printing and screen printing, both have reached a point where further innovation is required in the process of printing. From blocks and screens, it is now time to move on to something more customized, where people can contribute in the designing of the prints and impressions. Just like engraving, where people can customize utilities, hand block printing should be made more customization oriented.

Figure 7.6 Printed cotton kurta in beige and red

7.2.10　Wearing block prints

Experimenting with looks is so much fun with block prints. A long Flared Skirt with a Vest Top, complimented with delicate metallic jewelry gives a chic and modern look. For a traditional look, try a block printed salwar suit with light pearl jewelry.

Block prints are multi-ethnic. Their global and Indian demand stems from the fact that the print industry is an indispensable part of fashion and interiors. Climate is not a constraint for block printing because the fabrics used for block printing could be of any nature.

7.2.11　Maintenance

You don't have to worry about your block prints much. Some basics like washing them on the reverse, ironing them regularly and keeping them free of stains can easily add value to the life of the block printed fabrics.

7.3　Facts and comparisons

- Hand block printing will cost you five times less than a Phulkari and is five times more popular than it
- The state of Rajasthan is the home of hand block printing
- Hand block printing technique is used in almost every industry related to fashion and interiors
- Use of natural colors has increased the demand for hand block printing by more than 40%
- It is extremely difficult to find out whether a fabric is screen printed or block printed
- During the Mughal era, the block printing technique was a design technique much admired by the imperial class.

7.3.1　History of the block

Various tribes or groups of people, the world over have their own symbolism. A specific colour; a particular kind of embroidery, a distinguished style of jewellery or the patterning on their garments helps us recognize them. This gives them their own unique identity. Even today we find communities in remote areas in India as well as other parts of the world, where a specific style of printing in precisely the same design and colour is worn by every individual of the same group.

Today India has many major printing centres with their own block making skills and history. Wood blocks are largely used for printing fabrics for costumes, floor coverings, bedspreads and sometimes even wall hangings or prayer rugs. Blocks are also used to transfer designs, which were used by the embroiderer as a guideline for embroidery; for example in – Chikankaari of Lucknow, Kashmiri shawls and Pheran; embroidered yolks in Rajasthan and Gujarat. Another technique where blocks are used to print the basic design before the real work started was tie and dye or baandhani from Kutch.

Photo credit: Vikram Joshi

Figure 7.7 Block for printing Kashmir embroidery design
(Courtesy: Private collection of Vikram Joshi)

7.3.2   Customs and practices associated with blocks

A block printing unit which has been running from generations will probably have thousands of blocks. Blocks are rarely thrown away. One of the reasons for keeping the blocks is that there are many designs which come back in fashion after a few years.

Another way of documenting the design of each block is to take a "Parat", a precise impression of the fresh block taken by the block printer before printing starts. This remains in the records, as a future reference for the block maker.

Figure 7.8 Colour being applied on the block for the first time to make a Parat

The more organized printers have books, which will have a chaap, a print of each existing block in the unit. A particular number is assigned to each block and the same is mentioned on the print in the reference book. This catalogue is supposed to help one in finding the blocks. The catalogues are very useful for designers or business establishments, who want to work in these units. Another way of cataloguing- that we sometimes come across is filling different styles or collections of blocks in small cardboard boxes, and then sticking the block designs outside the box. These prints also mention the number assigned to the block.

Figure 7.9 Cataloguing blcoks with numbers

Old blocks are never sold by traditional printing families. Sometimes if the blocks are too old and completely useless then they might be taken to the river and given away in a proper ceremony, like the ashes of the dead is given to the rivers. This shows how alive the blocks are to the printers.

Most printers usually have memorized the design of each block in their mind, and the moment you describe one, they will not look at the catalogue, but just tell you where to find it. They would tell you exactly on which shelf, in which direction, or under which table one must be able to find it. Ironically many of them have never been to school, or if they have been there it was so brief that it is not worth mentioning. But that does not change the fact that without using a ruler they can make the straightest line that an artist can make after being trained in an art school. Their eyes and hands are so well trained you can actually see them measuring each border by the movements of the eyes.

While walking through various block printing units and getting work done there, one gets more and more familiar with the customs, which are quietly followed. One slowly learns about the strict training and the kind of education that the printers go through during the course of action. When a new person comes to learn how one of the old masters or a well trained person will take the pupil – Shaagird under his guidance. The master himself usually will be a practicing printer and very respectful towards the need of the establishment for whom they both work. He will not only take good care of the new pupil, but teach him from step one. The new comer watches the movements of the hand, while doing all the side jobs for the master or the Guru – the teacher so that he is not disturbed during the printing process. The ultimate goal is to do maximum good quality production within the limited time frame given to them as a team, while keeping a balance of the hierarchy.

Details of Hand block Printing Block making: The typical hand block print had no large, uniform areas of colour but was skillfully built up from many small coloured areas, because wooden surfaces larger than about 10 mm in width would not give an even print. This had the advantage that a motif such as a flower would have an effect of light shade obtained from three or four blocks, each printing a different depth of the same colour or shade. A fairly hard wood was required, such as pear wood, and four or five layers were usually glued together with the grain running in different directions. The design was traced on to the surface and a fine chisel used to cut away the nonprinting areas to a depth of perhaps 1 cm. To obtain more detail from some blocks, strips and pins of copper or brass (more usually) were hammered into the wood. In the 19th century some blocks were made with the printing surface entirely in brass, which gave very delicate prints. Another technique used for complex designs was to prepare a mould to cast the image from molten type metal, fasten the casting to the block, and then grind the surface perfectly flat. When large areas of solid colour were required, the areas within metal or wooden outlines were filled with felt, which would absorb and print the paste uniformly. Finally, each block required corner 'pitch pins' which printed small dots; these allowed the succeeding blocks to be correctly positioned by accurately locating the pitch pins above the already printed dots.

Different techniques of hand block printing: Discharge printing: In this technique before the printing is carried out first the fabric is dyed to desired colour, then the dye is to be removed at selected places by chlorine or other colour destroying chemicals (which yields a white pattern on a coloured ground) from the part of the fabric where the design is to be printed. Coloured patterns on a dyed ground are possible in this method by adding to the bleaching paste a dye not affected by the bleaching agent used, so that another colour is substituted for white on the dyed ground.

Direct block printing: The fabric is first bleached and then dyed with desired colour. After that the hand block printing is done with carved wooden blocks in borders and in the inside of the fabric

Resist printing: In resist printing the design desired, is printed on the fabric with a material (Wax or resin) which will resist dyeing. The fabric is then dyed with desired colour. Washing after dyeing removes the resist material in which design is printed leaving a white pattern like the following print effect is achieved on the fabric.

Rich and colourful prints can be created through block printing. In olden times it was done with natural dyes but now a days, it is done with artificial colours and synthetic dyes. The colours commonly used for printing are saffron, yellow, blue and red. The wooden blocks are used for printing. They are of different shapes and have designs carved at the bottom of the block. Teak wood is used for making them on which designs are made by skilled craftsman. These blocks are known as 'Bunta'. Every block consists of a wooden handle and 2-3 holes which are made for the purpose of free movement of air. The blocks before taken into use are kept in oil for 10-15 days, which provide them the required softness

Select the fabric batch for printing as instructed. • Wash the fabric selected and bleach if required as per instructions • If the borders are to be made, tie the cloth at the border area and take for dyeing • as instructed. When the fabric is ready for printing spread the fabric on the printing table and fix • firmly with small pins on the table While spreading ensure that the fabric is laid on the table uniformly and no • crease/hold is there. Keep the colour tray ready with required colours of proper mixing as instructed by • supervisor.

Dip the wooden block up to design level in the colour tray to transfer colour on the • design portion of the wooden block. Ensure to evenly immerse the design portion of hand blocks. • Ensure to avoid over inking or un evenly inked block. • Now take the ink immersed hand block on the fabric and apply (printing on fabric) • by pressing it hard on the cloth in an uniform manner. Always commence the process of block printing from left to right. • Repeat the printing by following the point marked on the block which facilitates • sequential order of printing on the fabric Ensure to make uniform repeats of design as instructed and avoid un evenly • placing of blocks on the fabric. Follow the instructions of supervisor and skilfully print to get uniform and clear • block printing motifs on the fabric. Single colour printing shall be done by one operative and takes less time for • printing. In case of multiple design / colour printing carry out printing with the help of • additional operatives. In block printing, it is not easier to bring colour variations. Bring colour variations • as instructed. If testing of colours is requirement for their fastness before applying on the fabric, • ensure to get the sample tested and the reports are approved by supervisor to proceed further. Confirm with supervisor which type of dye is to be used, as there are different • types of dyes are used for printing cotton fabric- such as indigo sol, pigment dyes and rapid fast dyes. In case of rapid dyes once prepared for printing have to be utilized on that day • itself and can not be stored, hence confirm with supervisor what is the quantity to be mixed for printing. The Pigment colours are commonly used as its procedure for usage is simpler • compared to other dyes. The other advantage is after mixing pigment colours for printing, it is not • necessary to use the mix on the same day, as they can be stored in containers having lids for some time. Before application for printing the pigment colours mix in right proportion with • kerosene and binder as instructed. It is the operator's responsibility to ensure about proper proportion of mixing as • per supervisor's instructions before applying for printing.

If new shades to be obtained by combination of basic colours confirm with • supervisor for different proportions to be followed. The drawbacks of pigment colours are that, they consists of tiny particles which • don't dissolve completely as a result, minute residues are left on the fabric, while applying pigment colours be extra cautious to avoid quality problems because of tiny particles. Indigo sols and rapid dyes normally get completely soaked into the cloth and work • as instructed for these dyes. Indigo sols do provide vibrant colours like pinks and greens. After printing dry the printed material in sun, to fix the dye to the cloth. • After drying, roll each layer of cloth in newspaper and steam them in boilers. • This procedure of steaming is to be done for all kind of fabrics. • After steaming process, take the fabric for washing· Follow the no. of washes as instructed by supervisor. • Once washing is completed take the fabric for drying in sun· Dry in sun shine for no. of hours as instructed by supervisor. • Finally iron the steamed, washed and dried fabric as instructed. • While ironing ensure to keep the folds as instructed. • This whole procedure fixes the print permanently on the material printed. Various • garments like saris, kurtas, shirts, salwar kameez, dupattas, skirts, etc are made from block printed fabrics.

7.4   Block printing introduction

Block printing is the art form which has turned out to be great source for fashion in India. People in metropolitan cities like to wear clothes with embellished block print designs. Stunning designs for apparels and furnishings are designed which has created its increased demand all over the world. And due to the different types of block printing in India, people are unable to experience a great variety of block printing techniques.

Block printing designs might differ according to the various places. The production is done on large scale at places like Rajasthan, Gujarat, West Bengal, Uttar Pradesh and Andhra Pradesh. You can see slight changes in the design of different regions. For instance in eastern regions of India, you will notice that the block prints are bold and large whereas if compare it with the block print designs of western region they are not so bold. Lepakshi and Ajarakh are the different types of block printing designs which are mostly demanded by people.

Particularly in Rajasthan, the variety of Rajasthani Block Prints which you can see is the Sanganeri print and Bagru print. As these designs have emerged from one state only, thus much difference cannot be seen in them. The difference lies only in the background color of both the prints. Mostly Sanganeri prints are designed on the white background fabric and Bagru print is created on the red or black background fabric. Due to different backgrounds the fabric gets different look and the motifs are emphasized differently.

In fact, there are diverse techniques which are used to create different types of block printing designs. The techniques which are applied by the crafts men are given below.

7.5   Types of block printing techniques

Discharge Printing: In discharge printing the complete fabric is dipped in color and then the particular area on which design is to be created. It is then reacted with a chemical to make it faded from that specific area and thus then the faded portion is colored again with different color. This block printing process leaves the fabric with special colored affect of block printing in textile.

Regular direct printing: This is the most common technique which is applied on the fabric to get block prints. The cotton or silk fabric is firstly processed under the bleaching process so as to increase its absorbing power for dye and then their grayish color is removed by giving them yellowish tinge. After this the cloths is left for drying and once it dries up, the block prints are created on the fabric by the wooden blocks and dyes. Firstly, the outline is created using crafted wooden block with black ink on it and then other details are applied with different colors.

Resist Printing: As the name of printing states, the color is applied with the resistant factor i.e some areas of the fabric are made resistant to color by applying a mixture of resin and clay on it. Then the color or dye is applied to rest of the fabric which leaves the fabric with undulate designs which are considered beautiful in block printing.

India is amongst one of the major exporter and manufacturer of block printed fabric across the globe. Hand block printing is the most ancient form of dying in which craftsman make use of wooden or metal block to create blocks which is also known as ''chappa''. Today, this technique is widely used to deliver the best Hand block printing services in India.

7.6   Techniques of hand block printing

There are some techniques which are used for block printing in India.

Figure 7.10

Direct Printing: In this form of dying silk or cotton fabric is bleached first and then dyed. Subsequently dyed fabric is then printed with blocks.

Resist Printing: In resist printing some areas are sheltered with a mixture of clay and resin to shield them from dying. The dyed cloth is then washed and color spread into protected area through splinter, producing an undulated effect; thereafter blocks are created for designing purpose.

Discharge Printing: In this technique, a chemical is used on dyed cloth to remove the dye from sections that are supposed to have further design and then blocks are printed.

7.7   Hand block printing in India

Bhavnagar, Porbandar, Kutch, Rajkot, and Jamnagar are famous for hand block printing services in Gujarat. A village in Gujarat known as Dhamadka is renowned for a block printed fabric named as Ajrakh, basically geometric design is popular for this using natural color likely black from tarnished iron and indigo is used for the blue color. Other popular patterns are sculptures of dancing girl, animal and birds.

In Rajasthan, Bagru is most famous centre for hand block printing services known as Syahi and Dabu prints. Apart from this sanganeri print is also very impressive for bedsheets, bed covers and coverlets. A style of printing known as Doo Rookhi is eye catching in which artisan stamp on both sides of fabric. Prints of horse, camels, peacock, lion, god and goddess are common prints using bold color and pattern.

Figure 7.11

Geometrical and floral prints of Punjab are attractive and colors used are basically pastel or light shades. This is an art of a group of textile workers known as Chhimba. In West Bengal, Serampur is known for its bold and vivacious patterns of printing. Andhra Pradesh is prominent for its Kalamkari printing which is a combination of block printing and hand painting. Sri Kalahasthi and Masulipatnam are major centers of Kalamkari. As the name suggests; a special type of brush cum pen is used for this type of printing.

In hand made printing stamp plays a very important role so it is imperative to make good stamps to make design more appealing and attractive. The process of creation of stamp usually takes 8 – 10 days depending upon the design. Majority of designs are inspired from Hindu mythology.

7.8   Process of block printing in India

Step 1: Carving blocks and dying

Blocks are carved with beautiful design with the use of hammers and drills and used for the purpose of stamping, Afterwards these wooden blocked are soaked in oil to prevent them from cracks and making them more effective and durable.

Step 2: Bleaching of fabric

Fabric to be printed undergoes with the process of bleaching to make the fabric soft and remove all the starch.

Step 3: Applying color paste

After carving these blocks are dipped into a mixture of colors and gum before they are stamped on fabric.

Step 4: Printing

Now, these colored dipped blocks are stamped on fabric to make the actual design, printing always to be followed from left to right across the length and breadth of cloth.

Handblockprint.com is the leading hand block printing services provider in India to buy the hand block printed sarees, suits, kurtis and more.. This whole process will help you to understand the clean and beautiful printing behind the ready-made printed garments

7.8.1   Pigment dyes

These colors are readily available in the market and are easy to use. The mixed colors can be stored in plastic buckets after use. Pigment colors, brought from the market are further mixed with kerosene and a binder. The mixing has to be done carefully as the thickness of the material can give raised effects on the cloth while printing. These colors follow the direct printing technique. Colors applied are visible and do not change after washing. A number of colors can be obtained by mixing two or more pigment colors.

7.8.2   Reactive dyes

These are the chemical dyes which when mixed with second chemical produces a third color. Artisans therefore dye the cloth, to be printed, in one chemical and then print it with another chemical. These two chemicals react with each other and hence produce a different color. There are only few chemical dyes available in the market.Colorant Reactive dyes are giving good results.

7.8.3   Rapid fast colors

These colors are difficult to store and has to be used the same day. In rapid fast color process the color in the design and the ground color both are printed in one go. Generally white or light background is used. There are only few colors available in this process.

7.8.4   Cloth

Traditionally the printing was done on white or pale background of cotton cloth. Today the craft is practiced on any material ranging from cotton, silk, organza, jute, kotadoriya, chiffon, paper etc.

7.8.5   Kambli (Woolen cloth)

A piece of woolen cloth is laid in the color tray. This helps in proper application of color on the block.

7.8.6   Drying

After the printing is completed the fabric is dried out in sun for the colors to get fixed. This is done specially for the pigment dyes. The printed fabrics are handled with utmost care so that the colors are not transferred to other areas. Therefore they are wrapped in plastic or newspaper after dying. The process is called as Sukhaai.

7.8.7   Washing

Fabric then goes through the process of steaming in the special boilers constructed for this purpose. After steaming, the material is washed thoroughly in large quantities of water and dried in the sun. Once the fabric is washed and dried ironing is done, which further fixes the color permanently. This final process of washing is called Dhulaai.

7.9   Process B

7.9.1   Block carving

Block printers, carvers and dyers are all traditionally part of the Chhippa cast in India. Like most crafts in India the skill of block carving is passed down from father to son. Open fronted shops dotted along the road are filled ceiling high with stacks of wood ready for carving. Chhippas squat at small tables with their traditional tool kit of miniature chisels, hammers and drills and begin carving intricate patterns into the wood.

7.9.2   Wood blocks

After each wood block is carved they are soaked in mustard oil for up to a week to ensure the wood doesn't crack when exposed to the dry conditions of the printing process.Tiny holes are also drilled through each block to ensure that the wood breathes, allowing the blocks to last for decades. Often when you go into printers studios you can find old blocks covered in dust and tucked under the printing tables, dating back 80 years, each holding their own history.

7.9.3   Mud paste

Dabu is a smooth paste, which combines well sieved and soaked black earth, tree gum and a powder from wheat grains. The printer gently pats the wood block onto the dabu paste then quickly stamps it onto the fabric. This paste acts as a resist during the dyeing process.

7.9.4   Block printing

The surface used for printing is a saree length table (approx 6 meters) that is padded with many layers of cloth. The printer aligns the first block to the bottom left corner of the fabric and with incredibly precise hand eye coordination, that has been developed over years of block printing, gives a sharp tap the release the dabu paste onto the cloth. This same process is repeated along the length and width of the fabric.

7.9.5   Dusting

A fine saw dust is scattered over the wet daub paste once it has been printed to prevent the design from smudging and seals the printed portion from the subsequent dyeing process.

7.9.6   Drying

Once the sawdust has been scattered, the fabric is taken outside so the sun can dry it. This causes the sawdust and dabu paste to fuse together, creating a hard barrier that the dye can not penetrate.

7.9.7   Indigo dyeing

The dabu printed cloth is next immersed in the deep vat of indigo dye. Natural indigo, from the indigo plant *indigofera tinctoria*, is not water soluble. It is purchased in blocks, ground into a powder and soaked before fermenting in an underground vat containing a strong alkaline lime powder and water. The strong alkaline reduces the indigo dye, removing oxygen from the liquid and so making the colour chemically available to bond to the cloth. When the cloth is removed from the vat it is green in colour, though as it comes into contact with the air (oxygen) the cloth develops into a rich blue tone. The cloth is dipped repeatedly into the indigo vat to achieve darker shades of blue, drying thoroughly between each successive immersion.

7.9.8   Drying fields

After the fabric has achieved the right shade of blue, the cloth is stretched out in the sun to dry. In Bagru you can often stumble across large fields in the desert that are covered with fabric, appearing like a giant patchwork quilt or one of Christo's latest installations.

7.9.9   Wash baths

Washing occurs at the start and end of the printing process. The fabric is first soaked in large outdoor water baths for up to a few days to remove any starch, oil, dust and any other impurities. Once all printing and dyeing procedures are completed, the cloth is once again subjected to washing and beating in the baths to remove all traces of the dabu mud, revealing the resist area to be the original white.

7.9.10   Indigo fabrics

Each piece of indigo fabric tells a story of where it comes from in that it's end colour can be influenced from anything from the weather conditions of the time at dyeing to the pH levels in the dye vat to the minerals in the water or the consistency of the dabu paste. However it is in these uncontrollable elements that the beauty of indigo dyed fabrics lies.

The traditional process of hand block printing on textiles, with rich natural colors, has been practiced in Rajasthan for around 500 years. Block printing was introduced to the Jaipur region of Rajasthan by the Chhipa community. This community was originally located in Bagru Village, an area now famous for its vegetable dye and mud resist (dabu) block prints. The art of block printing has been passed down for generations within families and communities and has branched out in recent decades to other regions such as Sanganer, just South of Jaipur. In traditional Bagru style block printing, the 'recipes' for the traditional plant-based dyes are developed within each family and kept alive from generation to generation. The colors are dependent on the quality of the plants, the water and skill and knowledge of the printing masters. In more recent forms of block printing, such as those practiced in Sanganer, colors are mixed using azo free pigment dyes.

7.10   Woodblock printing technique

Woodblock printing is one of the earliest printing techniques. It creates a natural, almost vintage effect on natural fabrics such as cotton, linen and silk.

William Morris used this method, to great effect, for some of his materials. For every colour used in the design, a separate wooden block must be carved. The larger, heavier designs are carved first while the more intricate detailing is left till last.

In some instances, a method, called coppering, in which strips of brass or copper are driven into the blocks to represent finer design details, is also used. The colour is applied to the block and pressed firmly onto the cloth. This process is then repeated until the full length of fabric is printed.

Advantages :
- It's a process that imparts a simple yet intriguing effect to the fabric.
- It is a simple way of transferring patterns and text to lengths of material.

Disadvantages:

- If a selection of colours is to be used, each one must dry thoroughly before the others are applied.
- While the effect of block printing is simplistic it is a very time-consuming method and could become quite an expensive endeavour once the carving time is considered.

Costs:

Fabric and dye costs are relatively affordable. The carving of the wood block can be labour intensive depending on the intricacy of the design. Therefore, the costs could be quite high.

7.11   Process

7.11.1   Block carving

A print starts with the design, drawn on paper and carved into the Sheesham wood block. Designs are meticulously carved by hand into the blocks which are approximately 18-25 cm across. The physical block is the design for a single repeat which is then stamped in rows across the fabric. Each colour in the design is carved into a separate block. The outline block or 'rekh', is the most intricate and usually stamped first; it is typically the outline for a floral or lattice type design. Next comes the fill block or 'datta' and possibly the ground colour block or 'gud' depending on the colour scheme used. Block carving is in itself an art requiring years of apprenticeship to gain mastery and is done entirely by hand.

7.11.2   Color mixing-preparing the dyes

Once the blocks are carved, the master printer prepares the colours which will be used in printing. The colours are then poured into wooden trays and the blocks stamped in the colour each time, then stamped onto the fabric to form the repeat pattern. The colours shown are azo free, eco-friendly synthetic colours which are used in Sanganer printing.

For each new design, we do a colour check and test out new colour combinations. We use Pantone TPG (TPX or TCX) reference codes for colour matching. Above, our master printer,

7.11.3   Printing process

Block printing is an ancient textile tradition that originated in the Rajasthan Desert region of India centuries ago. It involves stamping designs onto fabric by hand using intricately carved wooden blocks. Block printing is a labour-intensive, painstaking process which requires time, teamwork and a tremendous level of skill. For example, it can take five carvers up to three days to create a pattern in a printing block. And the printers may use as many as thirty blocks to complete one design!

Despite competition from faster and cheaper methods of textile design, block printing has resisted industrialization throughout the years and is still done without any mechanization. It has endured in popularity because of its natural feel and its ability to achieve a three-dimensional, artistic aura which is difficult to replicate in machine-made and mass-produced items.

7.11.4   The process

The three main tools of block printing are the wooden blocks, the fabric and the dye:

- Raw fabric is first washed to remove starch and then bleached in a gentle solution.
- Once bleached, it is hung up on tall bamboo frames to dry in the hot desert sun.
- The blocks are made from teak wood by trained craftsmen. Designs are traced onto the surface and then carved with a hammer and chisel.
- Each block is made with a wooden handle and several small holes to release air and excess dye. It is soaked in oil for 10-15 days before use to soften the grains of the timber.
- Colours are mixed in a separate room from the printing. They are then kept on a tray which the printer drags along as he works.
- In preparation for printing, the cotton fabric is stretched tightly over the printing tables and fastened with pins to avoid wrinkles and movement.
- When printing begins, artisans first print the outline of the design. This is usually done by the most experienced printer, as the outline leads the whole process and must be very precise.
- The rest of the team then fills in the colours with the various "fill" blocks.
- Once printed, the swaths of fabric are hung to dry in the sun before a final wash.

Each colour pattern is stamped individually onto the fabric; the process takes skill and time, as the pattern must be stamped repeatedly across the fabric, colour by colour. The slight human irregularities — inevitable in handwork – create the artistic effect emblematic of block prints. The final outcome of this intricate labour is a timeless beauty, and every garment made from this fabric is unique.

The printing master must carefully align each block as he prints, using the 'guide' carved on the left edge of the block as his marker. Each printer has a slightly different style which is considered his 'signature' look. The printing master must then follow the same pattern of aligning the blocks with each color layered on to the design. The subtle gaps and overlaps are a beautiful reminder of the hand work and give block printing it's iconic look. All prints exemplify this aesthetic and have a subtle pattern of light/dark across the design.

The block printing villages are know for their rhythmic 'tock-tock' sound of the block printer hitting the wood block to 'stamp' the pattern. It is an enchanting sound which echoes through the village and is a reminder of the significance of artisan work.

7.12   Style of printing

The original Bagru style printing traditionally used natural vegetable dyes and mud resist techniques to print on cottons and silks.

Traditional Bagru designs reflect nature in floral, leaf and geometric motifs. Later techniques incorporated Persian motifs and developed block printing into a highly intricate style.

Hand block printing now also extends to Jaipur and Sanganer with the use of azo-free, synthetic dyes (such as pigment and indigo sol) and different styles of printing.

7.12.1   Indo-Western designs

In recent decades, designers from the West have worked closely with local artisans to create Indo-Western styles which are inspired by other cultures, pop-art, nature and city-scapes. This collaboration has been beneficial for everyone as new designs emerge, but also it helps to tell the story of block printing and keep the market alive. Designers work both in Bagru and Sanganer. In Sanganer, few have their own facility using azo free, eco-friendly, pigment dyes with their own print designs. The advantage of 'modern' synthetic dyes is that they are colourfast, easier to make and machine washable. In Bagru, the designers work closely with a local printer using vegetable dye and dabu mud resist techniques in traditional prints such as indigo prints.

7.13   Bagru printing

7.13.1   Dabu printing and vegetable dyes

The "recipes" for Bagru style vegetable dye prints have been preserved for many generations by the artisans' families. Many of the dyes require months of curing for the desired colour to develop. Weather, water quality, and changes in the crops, all affect the vegetable dye.

7.14   Social impact

7.14.1   Decentralized artisan textile production

Block printing is typically done in open-air facilities in villages, or in people's homes. It provides a source of income to many village families and is an environmentally positive approach to textile production in rural India. It is also a method of decentralized production, following Gandhi's philosophy of keeping more people employed within their traditional environment. While often men have been the printing masters, in small-scale, traditional production, women also become skilled printers. Traditional printing is often done in family units which provides more income for the whole family and allows women to work within the the day-to-day routine of family life.

7.15   Artisan standard

Block prints, are, by nature, hand-done. The slight color variation within a print run and across different print runs if printed at different times or in different seasons, is a natural part of the process. It is an attribute appreciated by those who value the uniqueness of artisan textiles.

Block printing is an ancient textile tradition that originated in the Rajasthan Desert region of India centuries ago. It involves stamping designs onto fabric by hand using intricately carved wooden blocks. Block printing is a labour-intensive, painstaking process which requires time, teamwork and a tremendous level of skill. For example, it can take five carvers up to three days to create a pattern in a printing block. And the printers may use as many as thirty blocks to complete one design!

Despite competition from faster and cheaper methods of textile design, block printing has resisted industrialization throughout the years and is still done without any mechanization. It has endured in popularity because of its natural feel and its ability to achieve a three-dimensional, artistic aura which is difficult to replicate in machine-made and mass-produced items.

7.15.1   The process

The three main tools of block printing are the wooden blocks, the fabric and the dye:

- Raw fabric is first washed to remove starch and then bleached in a gentle solution.
- Once bleached, it is hung up on tall bamboo frames to dry in the hot desert sun.
- The blocks are made from teak wood by trained craftsmen. Designs are traced onto the surface and then carved with a hammer and chisel.
- Each block is made with a wooden handle and several small holes to release air and excess dye. It is soaked in oil for 10-15 days before use to soften the grains of the timber.
- Colours are mixed in a separate room from the printing. They are then kept on a tray which the printer drags along as he works.
- In preparation for printing, the cotton fabric is stretched tightly over the printing tables and fastened with pins to avoid wrinkles and movement.
- When printing begins, artisans first print the outline of the design. This is usually done by the most experienced printer, as the outline leads the whole process and must be very precise.
- The rest of the team then fills in the colours with the various "fill" blocks.
- Once printed, the swaths of fabric are hung to dry in the sun before a final wash.

8

Digital textile printing

Textile industry has taken a big leap in the digital printing sector. One of the most promising developments in the textile industry is digital fabric printing. It has opened the doors for numerous prospects to enhance the quality and maintain the growing demands of textile printing. Anything can be printed with ease and perfection on fabric using digital printing technology. Finally the digital revolution in textile ink jet printing has started .Globally more than 20 million meters of textiles are printed. Mainly due to high cost production more than 50% of textile printing is processed in Asia. Rotary screen printing has by far the highest market share, followed by flat bet printing. In textile printing, in recent years there has been special interest in digital printing. However, digital printing is still comparably small with less than 2% market share, substantially less than in other non-textile printing applications. The textile printing market is undergoing a transition from conventional printing (rotary screen or flat bet printing) to digital printing (ink jet printing), which will enable manufacturers in the textile processing industry to shift from mass scale long-run to on-demand manufacturing, responding to the increasing demands for fast fashion by short production runs and customisation.

Current growth rates (CAGR) in this industry are approximately 20%, but the overall market penetration of digital prints in printed textiles is still low – below 3%. In the European market the penetration is higher due to trend of small run lengths in a high cost environment and ecological concerns which favour digital printing over conventional printing.

We all know in 2014/15 some key mergers and acquisitions took place in the digital textile printing sector. Stepchange Innovations GmbH has been involved in some of these transactions as technology advisor in the due diligence process.

Private equity financial investors as well as strategic investors and large US corporations have entered the technology driven European digital printing space.

Generally, the mergers create companies with the market reach and financial power of large global corporations blended with the technical capabilities and know-how of specialist companies.

8.1   Why digital printing?

Digital (ink jet) printing combines many advantages: easy coloration, direct printing off the computer screen, no screens, no colour kitchen, and attractive designs – even designs which are not possible with any other printing technique.

In simple words, when digital images are reproduced on physical surface, it is called digital printing. The physical surface can be in forms like paper, cloth, plastic, film, etc. In digital textile printing, once the design is created, it can be directly printed on the fabric from the computer. This process does not require any other step. Just as an image is printed on the paper easily, one can print designs on fabric.

Digital inkjet printing of textiles opens doors to new opportunities and creates new markets. Creative designs can be digitally printed that cannot be screen-printed. The largest screen printers have no more than 12 screens, which equates to a limitation of 12 spot colours. With process colour there can be an almost unlimited number of colours in a design, allowing much more than 12 colours in a specific design. Design cycle times are reduced and sample production can be done immediately. The ability to do economical short runs allows reductions in the size of inventories. Restocking of a 'hot› apparel item is made easy by digital printing and the store doesn›t have to discount its prices. Today›s markets are changing faster and customers are becoming more demanding than ever. Digital textile printing allows the production of goods and services to match individual customers› needs.

To print the designs on the fabric digitally, a dye-sublimation printer is used, which carries out the printing process by using heat to transfer design onto the fabric. In digital printing, it is necessary to pre-treat the fabric. This will ensure that the fabric holds the ink well, and a variety of colours can be attained through the pre-treatment process.

Digital textile printing is considered to be the 'next generation' printing which is quite different from the conventional fabric printing. With fabric printing going digital, many textile entrepreneurs are coming forward to invest in digital printing technology, as it is the most budding method of printing. In India, the textile industry is embracing digital printing technology by printing novel designs on saris and dress materials to meet the demands of domestic and international markets.

The textile industry in India has advanced tremendously in last ten years. According to the industry sources, digital printing in India caters to 1% of the global demand of printed textiles. However, in the coming five years this share is going to increase to 10%.

With digital printing gaining huge popularity and technological improvements that have taken place in ink, consumables, print heads and printing machinery, the global production of printed textiles had reached 32 billion square meters by 2015, according to a global report released in U.S.

According to the fashion gurus, it was predicted in 2013 that the future will see digital prints ruling the fashion arena. The past trends will be revived, with digital prints enhancing it further. That same year it was demonstrated variety of abstract digital prints that will uplift the overall look of the garment with its colours and designs. The fashion weeks that were held that time in Paris, Milan, London and other cities had showcased this trend.

Digital prints are for people who like to experiment with bold and innovative designs. The leaders of digital fabric printing like Alexander McQueen, Mary Katrantzou and Erdem Moralioglu have displayed a range of digital prints, varying from abstract prints to graphic paintings, which will stimulate one›s artistic senses. Digital prints have become the hottest pick in the fashion world today.

The dyes used for digital textile printing are different from the dyes used in traditional printing. The most popular dyes for digital printing are acid dye, reactive dye and disperse dye. These dyes can be used for all commercial applications, and have properties like bright colours, low to medium salt content and high colour fastness.

Various dyes are used depending on the fabric for digital printing, like acid inks are used on silk and nylon; disperse inks on polyesters and reactive inks are used on all cellulose based fabrics such as cotton, linen and rayon. Moreover, these dyes are compatible with low cost ink systems.

Digital textile printing is described as any ink jet based method of printing colorants onto fabric. Most notably, digital textile printing is referred to when identifying either printing smaller designs onto garments (T-shirts, dresses, promotional wear; abbreviated as DTG, which stands for Direct to Garment) and printing larger designs onto large format rolls of textile. The latter is a growing trend in visual communication, where advertisement and corporate branding is printed onto polyester media. Examples are: flags, banners, signs, retail graphics.

Digital fabric printing is act of printing digital files onto fabric using ink jet based printers.

Digital printing on fabric is a new and innovative process that involves printing a design, a pattern or an image directly from the computer onto the desired media by way of a large format digital printing machine, aka an ink-jet printer.

Digital textile printing started in the late 1980s as a possible replacement for analog screen printing. With the development of a dye-sublimation printer in the early 1990s, it became possible to print with low energy sublimation inks and high energy disperse direct inks directly onto textile media, as opposed to print dye-sublimation inks on a transfer paper and, in a separate process using a heat press, transfer it to the fabric. Digital textile printing is the latest innovation in textile printing. It is an inkjet-based method which allows manufacturers to print almost any design on virtually any fabric. The fabric will be pretreated, after which it will pass through the inkjet printer at high speed and will subsequently be dried, steamed, washed and finished. The digital textile printer uses a printable design of a data file, reads the right colour information and prints the desired colour onto the fabric with minuscule droplets of ink at an astonishing quick production rate.

Further advantages of digital transfer printing on textile:

- Transfer printing guarantees exact and unlimited colour match
- Recreated graphic details with unsurpassed accuracy on your fabric
- Easy and safe creation of complex patterns and design on paper
- Prompt response to market demand and production of short-run reorders with storage of your designs
- Feel free to choose high-end fabric for your next collection – we polish and perfect your design on paper before rolling out your fabric

•Is digital direct printing faster than digital transfer printing?

Yes, the sublimation process itself is faster when you leave out the transfer paper process. If your fabric has been adequately pre-treated and is ready to absorb the sublimation ink, the result may be quite satisfactory. For some types of fabric, i.e. polyester which is our preferred material, we always recommend transfer printing as the disperse colours risk setting off and detailed pattern designs can be difficult to achieve.

Subsequent to direct printing, your fabric is likely to need thorough finishing in terms of special washing etc. to obtain light and colour fastness. Not all textile-printing companies have the facilities to offer this finishing in-house. Such sub suppliers may increase the total printing costs and prolong the delivery time.

Make transfer printing your preferred printing technique for all your high-end polyester projects

Our customers handle many delicate textiles and the print is often essential to the final product. It goes without saying that fashion and home textile producers have strict demands for colours and patterns, but also our customers within medical textiles call for accuracy and durable prints.

Medical textiles and other technical textiles are often parts of a final product, i.e. a prosthesis insert or a medical instrument. The accuracy of the print is very important to fit other parts of the final product. Furthermore, these types of textiles have to meet all health and safety regulations.

It is important for all of our customers to waste as little fabric as possible. Developing textile projects by means of advanced software and paper is very cost-effective and the best way to a unique and beautiful result.

8.1.1   Test your textile innovation projects on paper

The digital printer and its transfer paper are important work tools in all innovation projects at scan htp a/s. Designers and technicians may unite their know-how and experiment with software, ink and transfer paper during the main part of their innovation journey.

Somewhere down the line, the innovation team will implicate the sublimation process in the project, but the extensive development work from drawing board to transfer paper saves many metres of expensive fabric and leads the way to the perfect result.

8.1.2   Digital printing versus conventional printing techniques

Like many others, we fell in love with the digital printing technology 15 years ago. Digital printing is perfect for print on polyester, which is our niche. In addition, digital print meets an increasing demand for sustainable production. The entire process saves resources and has a significant lower impact on the environment than any of the conventional printing techniques.

scan htp is living in a clover with the gigantic innovation potential of the digital printing technology.

8.1.3   More shades and photographic details with digital print

The digital printing technique gives us full control of the colours. We will match the colours in any original object, and this goes for delicate details in graphic patterns as well as photographs. The print translates accurately.

We still have to give the transfer technique some of the credit for the accuracy of our digital prints. Using transfer, we reduce the consumption of colorants and therefore, we avoid the risk of colours setting off. However, the gift of colours and shades we get from the digital printing technique is unsurpassed. In fact, only the digital printer's performance may set the limits of our choice of the colours.

We do use multiple colour patterns in conventional textile printing, but with a separate screen/stencil for each colour. The preparation in both gravure print and screen print takes time and is quite cost-intensive.

Do you need 10 metres, 500 or 5000? Digital print has removed the minimum limit

They say that the fashion industry dominates the market, but actually, it is less abstract than that. We decide ourselves what we want, and we have become much more conscious of our demands and perhaps even a bit picky. We want to stand out from the crowd at all times; the way we dress, exercise or decorate our homes. We think it is safe to say that the fashion industry is just playing along.

However we twist the consumption pattern, it is a fact that collections change faster than ever. With our digital printing technology, we welcome changing collections. Start-up costs are low once we have tailored the print file.

You may order or reorder the exact metres of printed fabric that you need unlike conventional techniques, which require orders of significant size to ensure a cost-effective production. We even order extra metres to avoid the risk of running short. Conventional printing techniques induce waste, cost money and put unnecessary strain on the environment.

8.1.4   No more worries about overlaps

Everyone that has worked with conventional printing techniques know that overlaps may be quite a challenge. To overcome it and create a satisfactory result where colours meet elegantly, accuracy and high quality machinery is essential.

Digital printing has eliminated this challenge because we control the print nozzles 100% via the software.

8.1.5   No limits of repeat size

In gravure printing the repeat size (length of pattern) is limited by the diameter of the cylinder (normally 60 cm). Screen printing is a little more flexible, but in most productions the circumference of the screen defines the repeat size.

With the digital printing technique, we have no limits of lengths whatsoever. This is probably the most ground-breaking features of digital printing on textile. Our customers welcome the new opportunities to design unique collections.

In our home textile segment, we have assisted several customers in changing from conventional prints to digital. We recreate the collection in question with the digital tools and provide the customer with an immediate improved result, but more so, the digital technique gives the customer an important flexibility for future collections.

Flexible repeat sizes also present a great advantage for fashion fabric, as prints for graduated sizes are no longer a problem.

8.1.6   The digital printer encourage innovation projects

Software, printer and paper are our essential work tools when we create new products or new functionalities. We have everything at hand to adjust shades, colorant supply, speed and many other issues. Testing, testing, and testing. And voila we reach a satisfactory result.

The digital printing technology is a true gift for those textile designers and textile technicians who want to transcend boundaries.

8.1.7   Digital print is slower

Yes, we cannot sidestep that fact. Conventional machinery may print up to 200 metres per minute (gravure: 150-200 metres/min – screen: 50 metres/min), whereas the digital printer performance is only 50-100 metres per hour. On the other hand, we do not spend work hours engraving or preparing screens. Digital printing gives you the exact number of required metres with an easy option for more if you should need it later.

The digital technology evolves all the time, also when it comes to print speed, but for now we have to justify the conventional printing techniques for very large orders due to the high speed and the relatively low price per meter.

8.1.8   Less impact on the environment with digital textile print

We do not discharge wastewater at all from our digital print production. In fact, our water consumption is very low. Colorants contain no solvents and are sprayed onto the transfer paper in a thin layer only. During the sublimation process, the polyester fibres absorb the colorants completely, and the transfer paper may even be recycled afterwards.

Our digital machinery has a modest energy consumption and therefore limits the CO2 liability.

Conventional printing techniques require adding solvents to the colour paste. Moreover, conventional printing requires quite a large amount of colorants to obtain a nice print. Cleaning of screens and machinery produces a lot of wastewater containing surplus colorants.

There are mainly two successful digital printing methods:

1. Sublimation Digital Printing
2. Direct Digital Printing

The technology behind both printers is same, but the media that is to be fed into the printing machine and the processes to get the fabric ready differs in both methods.

8.2   How sublimation digital printing works?

Sublimation has been used for several decades to apply images to upholstery, curtains, lampshades, carpets, cushions, furniture, window blinds and much more. Due to high set-up costs and the limitations of traditional print processes, it is only since the mid 90s and the advent of digital dye sublimation that high quality photographic images have been printed onto everyday items. Momentum has gained steadily and it is now considered a 'must have' service. The enormous range of substrates available continues to expand. Custom designs were simply impossible to produce before wide format inkjet sublimation. That's true of other areas of digital printing, but none offer the amazing range of items that can be imaged so effectively, and from just a piece of paper.

Digital sublimation has been used by many leading high-end fashion houses during the last decade or so. There are incredible polyester fabrics for designers to choose from and many feel far from synthetic. From glossy satins for ties, tops and dresses, to hefty thick canvases for luggage. For years, we printed fabrics with flat colours, or at best, coarse halftone images. Colours were mostly specified as 'spot colours' and CMYK (cyan, magenta, yellow, black) was rarely used. Having the widest colour pallet available for digital dye sublimation printing is certainly of great importance to the fashion industry where there is often less room for compromise, and digital print processes are being pushed to the limits of what colours are achievable. The sharpness and clarity of prints and breath-taking photographic imagery makes digital dye sublimation the only choice for many types of apparel, from performance sports to Paris couture.

Without a doubt, dye sublimation printing is one of the most effective methods for creating a whole range of customised and personalised products on-demand. This means you can deliver an almost limitless variety of creative and profitable applications, that will enable you to extend your range of services to your existing clients, or indeed open up a spectrum of new markets to you. With inkjet dye sublimation printing, you can print onto a vast range of fabrics and coated surfaces in many widths and lengths, including stretchy sports fabrics, heavy canvas, fire retardants and ultra-light voiles. In addition to these textiles, you can sublimate onto a range of coated hard surface materials, such as wood, metal, plastics, glass and ceramic. This means you can deliver a huge range of applications including sportswear and sports equipment, fashion, soft signage, interior décor, promotional merchandise and gifts. It's easy to learn, quick to deliver and means you'll continue to surprise and delight your customers. Extend your capabilities, satisfy their demands and beat your competitors by adopting digital dye sublimation printing.

We are not discussing about analogue 'traditional' print processes that run dye sublimation production, such as screen print, litho and gravure. Instead, our focus on dye sublimation with digital inkjet. Put simply, inkjet dye sublimation printing is the process where specially manufactured inks, containing heat activated 'dye sublimation' dyes, are fixed by heat and pressure into a polyester substrate, such as a polyester fabric or polyester coated surface.

Here are the key steps to be followed with dye sublimation:

STEP 1: Special heat activated inks (dye sublimation inks) are printed onto a digital transfer paper, usually as a mirror image.

STEP 2: Next we need a heat press (suitable for loose pieces) or a calender (suitable for continuous media on rolls) and a receiver substrate to apply the image to the substrate. The receiver substrate will be a polyester fabric or a material which has been pre-coated with a polyester surface, including glass, metal, wood, plastics and ceramics. The paper is placed on top of the item to be imaged, with the printed side down. Pressure and heat is then applied using the heat press or calender (typically 180-200°C for 35-60 seconds).

STEP 3: The print is complete when the paper is removed. No drying time, or post treatment is required. The process permanently images the polyester fabric and the print cannot be scratched or washed out. It becomes part of the fabric and has no handle, feel or texture above that of the fabric. This means it won't crack, peel or flake, and can be washed and ironed with minimal loss of colour. In the case of a solid object, such as a glass cutting board, the print appears beneath the coated surface and is very difficult to scratch or remove.

So why do we need to print onto polyester with dye sublimation printing? Polyester is a plastic, so when enough heat is applied, it begins to melt, allowing the pores to open up. When heated, the dye sublimation inks change from a solid to gas, and this gas can then enter into the open pores. This means that it is possible for the ink to be transferred to the polyester fabric or the polyester coated hard surface. This is not possible for materials such as cotton, paper, wood or wool which scorch and burn when exposed to temperatures required to activate dye sublimation inks (typically 180-200°C). That is why it is limited to just 100% polyester fabrics. As we have already mentioned, there are many products and materials which are coated with polyester, and they look stunning when printed. The opportunities are extended further through the availability of polyester sprays, allowing you to coat products yourself.

When compared to traditional printing processes digital wide format dye sublimation benefits from significantly reduced set up times and as such can be truly adaptable and flexible in its output. It's really easy to do one-offs, sampling or limited editions. Plus, you can also print different colour variations or completely different designs within the same print run. This means you can achieve mass customisation of one piece or you can batch up different jobs during one long print run. Either way it's maximum flexibility or maximum efficiency. Imagine printing fashion items such as dresses, bags and shoes and offering your customer these customised and personalised options. But what about high production work? Of course, you have the option of purchasing a separate higher-speed solution purely dedicated to high production. But, customers often choose a digital inkjet dye sublimation printer to sit alongside their production printer for sampling purposes. Alternatively, purchasing two or more printers can be another option to maximise production capability, whilst retaining the flexibility that digital print delivers.

When choosing a dye sublimation system, a typical question to ask is whether to opt for a wide format printer or a desktop solution. Desktop sublimation is sometimes seen as an easier route to take, but you have to ask yourself whether this will be a sustainable option for your business in the long-term. Desktop systems use A4 or A3 paper and are, therefore, restrictive in terms of the size of item they can deliver, plus the ink costs are typically very high. In fact between 5 to 20 times the cost of wide format inks. If you need to print higher volumes, then wide format printers are generally considered a much better option. In terms of productivity, wide format printers are simply able to print much wider and longer. This means that not only can you print hundreds and thousands of small items such as mugs, caps, t-shirts, bags, drink coasters, bar runners and mobile phone covers, but you can also diversify into larger items such as soft signage, sportswear, home décor and fashion. This ability to produce anything from a lapel badge to tear-drop flags offers a very compelling marketing story for any graphics producer. We might try and compartmentalise the services we offer by the technology we possess, but customers don't. The same person that buys 4m x 1m in-store washable fabric banners may also need shoulder bags, mugs or even guitar pics.

There are some wide format printers dedicated to printing directly onto fabric. This works well if you want a long run of one type of fabric. However, this does not offer the same flexibility that printing dye sublimation onto paper provides, because you can take a roll of printed paper and apply sections onto numerous fabrics or solid surfaces with a minimum of 75% polyester. Polyester fabric for direct printing has to be pre-treated with an inkjet receptive coating. This is usually done by the manufacturer or fabric wholesaler, so the cost is generally higher than uncoated polyester for heat transfer of inks from paper. Also, if there is an issue during printing, such as nozzle drop-out or feed/ tension issues, the cost of wasted fabric is considerably greater than the loss of equivalent paper. It is also worth being aware that it is still necessary after printing directly on polyester to heat press the fabric in order for the ink to fix onto the fabric and activate the final colour. It is possible to direct print other types of media, such as cotton, silk and wool. These fabrics must be pre-treated and require different inks, according to the type of fabric.

Ok, we've printed our dye sublimation inks onto paper – what type of heat press do I need to transfer the image onto my substrate? Calender presses are used to heat transfer from a printed roll of paper onto polyester fabric. The roll of paper is attached to the press and fed through with the fabric. Pressure is applied around a heated cylinder (usually around 180-200°C) to ensure even transfer and no creasing. Calender presses are generally available in widths of around 1 to 5 metres. Some are designed to enable pre-cut fabric pieces to be fed through individually, such as the front of a shirt or dress - pre-cut to shape for all-over print. Flatbed presses are used to print solid objects such as sheet metal, wood, ceramic tiles, floor mats, carpet tiles and small pre-cut fabric pieces. A popular addition to any dye sublimation solution is a T-shirt 'clam' type flatbed press. This is convenient for producing small fabric colour proofs before printing the full production. These can also be used to print small coated items such as tiles, metal signs and plaques. Smaller flatbeds vary in sizes, styles, performance and cost. In addition, specialist presses are also available for mugs, caps and pockets. Regarding settings, any heat press, paper, fabric or solid substrate combination will have its ideal temperature, dwell time and pressure settings to obtain optimum results. For example, a coated ceramic tile may require less pressure, but higher temperature and much longer dwell time than a fabric banner.

There are a variety of affordable heat presses that allow a paper print to image around 3D objects. Examples of such objects include phone covers, mugs, ceramic plates and drinking glasses. These desktop size presses use a vacuum to create negative pressure, sucking the printed dye sublimation paper around the object as heat is applied. In the case of smart phone covers, the paper is fixed using a heat resistant tape over a custom metal jig and placed in the 3D vacuum heat press. There is a growing range of suitable items to print on, and no post finishing is required – a key feature of dye sublimation printing. Of course, there are more industrial style ovens for larger items. Indeed, some businesses have developed specialist inhouse heat press equipment for bespoke items such as ski poles, bowling balls and even garage doors.

You need specific papers for the dye sublimation heat transfer process. You cannot use a roll of inkjet paper which is typically used for printing graphics or posters. Papers are specifically developed for dye sublimation printing. We do not want the ink to penetrate the fibres of the paper. We want the ink to stay on the paper after printing, but for as much ink as possible to be sublimated off the paper at the required temperature, not pressed back into the paper. A good paper from 90gsm to 140gsm will hold the same amount of ink. Heavier paper 'cockles' (ripples) less under heavy ink loading. Lighter weight paper is sometimes chosen, as it is cheaper and can provide adequate results. You can also run presses a bit faster with lighter paper. If you do a print run in advance and plan on transferring them days or maybe weeks later, the prints should be kept in a polythene bag (such as the bag the roll of paper is protected in). This will minimise absorption of moisture and maintain print integrity before heat transfer.

The last stage in dye sublimation is the finishing process. The extent to which finishing is required will depend entirely on the application and the substrate. For example, some applications don't require any finishing at all, such as mugs, floor tiles or mats, which once pressed, are ready for despatch. Other applications such as sublimated fabrics for sportswear or fashion require more complex finishing processes. It is not always necessary to bring finishing skills in-house, as there are many companies who can complete them on your behalf. This is useful for businesses who are just starting in sublimation, or only produce small numbers of items, and therefore don't want to invest in additional equipment or new staff. Should you choose to bring your finishing in-house, there are a number of options available, depending on which kind of applications and services you are offering to your customers. Here are some examples: soft signage: • simple cutting devices or a handheld hot knife will allow you to achieve a basic level of finishing for fabrics. A hot knife may be enough to stop fraying and unravelling, avoiding the need for sewing or welding; • for stretch frames, a silicone beading is stitched into the fabric. clothing, textile items and upholstery: • laser cutting machines and professional sewing machines offer a great deal of versatility, and the skill required to operate them is well within the capability of every print company. sublimated photographs on rigid materials: • photographs can be easily framed after printing to add value to the finished item. Finishing in-house offers considerable opportunities to diversify your business profitably and should be embraced rather than avoided.

From soccer, basketball and ice hockey shirts, to Lycra based fabrics for swimming, athletics and cycling, if they have graphics on them they have almost certainly been applied by sublimation. Most performance fabrics today are polyesters, and to maintain the fabric's feel and integrity, nothing does it better than dye sublimation. The vivid colours are also a prominent feature of dye sublimation inks. Performance fabrics have been developed to evaporate sweat to the surface and are extremely durable, yet stretchy, light and in an abundant range of finishes. Digital dye sublimation can just as easily produce a one-off or a short run of items for a team, or use the same overall image but customise each individual shirt. In fact, many designs can only be produced by inkjet sublimation, creating a completely new market for digital only sports clothing producers. Digital dye sub is also ideal for hard-wearing and eye-catching sports accessories and equipment such as gloves, helmets, skis, surfboards and much more.

Brands are increasingly choosing fabrics over traditional PVC based graphics. As such, soft signage is now a key application for print providers to consider. What are the reasons behind this trend? One reason is environmental. Fabrics will break down faster and can be recycled. Another reason is aesthetics. Vibrant white polyester is very striking and the printed dots blend in with the weave of the fabric, giving a continuous tonal range. For retailers, translucent stretched fabrics are a subtle alternative to traditional banners and important product placements are visible through the fabric. This translucency is hugely beneficial for exhibition stands too. Solid walls create blind spots, limiting the visitor's view of the stand messaging and products. Translucent fabric panels create the impression of structure without obscuring view and offer a more inviting feel. Polyester banners of several metres can be folded easily into a small courier envelope and shipped to customers without risk of damage. They are also very easy to handle at the customer's location, washable and easily mounted. Significant cost savings can be made on installation, as modern frames and sign systems can be fitted quickly and easily. For external short-term use (around 6 months) there should be little or no fading from all but the harshest climates. Internally, the prints will retain their integrity for many years. Floor mats will withstand muddy boots, stilettos and even jet washing, if the material can resist it, so can the print

There are plenty of pre-coated rigid substrates to choose from. As with fabrics, dye sublimation only works on polyester, so this pre-coating is a heat resistant polyester lacquer or powder coating. Polyester is a category of polymers. These polymers are often used in the production of smart phone cases, skis and snowboards and, therefore, can be suitable for dye sublimation. You can also choose to coat your own items. Most users will spray apply this and often a relatively low temperature oven bake will ensure firm adhesion to the surface. When imaging onto pre-coated wood, ceramics, slates and the large array of glossy, satin and matt finished metal sheets, the end result will often look like it's had a post clear lacquer applied. This is because the gaseous dyes have penetrated the polyester coating, creating a wonderful encapsulated finish.

Step 1> Ready Your Design:

RIP your digital print pattern, and get it ready for printing.

Step 2> Printing:

Print mirrored reflection of digital file on specially coated paper. Your print emerge as the print head passes right to left and back again across the paper, laying the ink on the paper in layers over a number of passes.

8.2.1   Sublimation digital printer

Step 3> Fusing / Heat Transfer:

Printed coated paper is then transferred onto any polyester-based fabric under extreme heat and pressure using heat transfer machine.

Few things to know here:

- The coated paper is also known as heat transfer sublimation paper.
- The process of transferring the print on fabric under extreme heat and pressure is known as fusing method.
- The heat transferred machine could be either oil-based or electric.

8.3   How direct digital printing works?

Step 1> Pre Treatment:

Apply coating chemical to the fabric before printing. This is also known as padding. Padding is useful because coating chemical holds and penetrates the reactive dyes better.

Step 2> Ready Your Design:

Choose the profile according to the fabric, and RIP the design to get it ready to print.

Step 3> Printing:

Your fabric is placed flat on the sticky printer belt. Coated fabric is then fed into the direct digital printer for printing. The fabric is then gets printed. This is the bit where we look a like a cat watching the tennis

Step 4> Drying:

Your printed fabric gets dried by the heater attached with the digital printer.

Step 5> Post Treatment:

The printed fabric is gone through the streaming, washing, drying, and ironing. This is the most vulnerable process. Each fabric requires a different length and quantity of steam depending on a number of variables including fabric type and length of fabric.

Digital fabric printing is a relatively new technology with tons of applications. I just completed my first line of digitally printed fabric earlier this year and I'd like to share a little bit about the technology, design process and possibilities.

8.3.1   Technology

Most commercially available fabric is rotary screen printed; each print run is typically several thousand yards. The high minimums are due to the cost and time required to prepare a unique set of screens, with each color in a design requiring a separate screen. The main advantage of digital printing is the ability to do very small runs of each design (even less than 1 yard) because there are no screens to prepare.

The inkjet printing technology used in digital printing was first patented in 1968. In the 1990s, inkjet printers became widely available for paper printing applications – you might even have one on your desk right now! The technology has continued to develop and there are now specialized wide-format printers which can handle a variety of substrates – everything from paper to canvas to vinyl, and of course, fabric.

The inks used in digital printing are formulated specifically for each type of fiber (cotton, silk, polyester, nylon, etc). During the printing process, the fabric is fed through the printer using rollers and ink is applied to the surface in the form of thousands of tiny droplets. The fabric is then finished using heat and/or steam to cure the ink (some inks also require washing and drying).

Digitally printed fabric will wash and wear the same as any other fabric, although with some types of ink you may see some initial fading in the first wash.

Figure 8.1 Photo courtesy: First 2 print

8.3.2   Design process

Designs can be created digitally with almost any graphic design software (Photoshop and Illustrator are the most popular). Alternatively, existing artwork or photographs can be scanned and then digitally manipulated to make a pattern. Usually designs are created as a seamless pattern that is repeated (tiled) across the fabric. You can also create a design that fills an entire yard without repeating, but you may run into issues if the size of the file is too large for the printing service to process.

Some helpful things to remember when designing for fabric:

Make color easy. Find out what color model your printer uses (most often CYMK or Lab) and choose your colors accordingly. You should expect colors to appear differently on the fabric than on your computer screen. Some colors such as deep, rich reds may be hard to reproduce. Large areas of solid color may come out with bands of lighter and darker tones. Setting up your design so that the colors can easily be changed (using layers or vector artwork) will save you a lot of headaches.

Focus on the finish. It's easy to get caught up in the artistic aspect of creating a beautiful design and lose sight of the fact that fabric is never the end product – it's always a part of something else. Make a habit of picturing the print as part of the finished product, especially concerning the size of the print. I have a ruler next to my computer – whenever I can't quite decide if the scale is correct, I'll hold the ruler up to the screen and zoom in or out until the size matches up. Sounds silly, but it works!

Print swatches. The color and texture of the fabric can have a noticeable effect on the print. Shiny fabrics like silk reflect light and can make the print seem lighter – thin fabric can be translucent and this will make print look washed out. Most digital printing services offer affordable swatches – even if they only sell by the yard, you can gang up a couple of designs onto a single yard.

Stay original. It may seem like a good idea to use digital printing to make a copy of a popular commercial print that is no longer available, but unlike clothing designs, print designs can be (and usually are) copyrighted by the artist or the manufacturer. It's best to stick with your own unique designs – if you're not artistically inclined, you can always hire a designer to make the perfect print for you.

Figure 8.2 Photo courtesy: Kayanna Nelson

8.3.3   Possibilities

Personalization. Every yard you print can be completely customized and personalized. Print fabric with names and dates, for use as quilt blocks, t-shirts, doggy raincoats, pillowcases, etc. Every item in your line can come in a different color. You can also do more practical kinds of customization, like creating sequentially numbered labels.

Lean and Just in Time (JIT) manufacturing. Small runs of fabric can easily be printed for sampling purposes. No more hunting high and low for the perfect print, only to find that it's no longer available when you need more. No such thing as fabric inventory – with an on-site printer, it's possible to print fabric on the same day that it will be cut. You can even have a pattern printed directly on the fabric.

Trends and fast fashion. Traditional screen printed fabrics usually appear on the market about a year after they are designed. With digital printing, it's possible to go from the design stage to finished fabric in a matter of weeks (or days, with an on-site printer). This allows you to take advantage of current trends, and even change prints or colors mid-season.

Figure 8.3 Photo courtesy: Lauren Spencer Hunt

8.3.4   Costs

The major downside to digital printing is the cost. As with any new technology, the costs are always high when it first becomes available. As time goes on and the technology continues to develop it will undoubtedly become more affordable.

When dealing with printing services, expect to pay around $20 to $40 per yard of finished fabric. Most services have no minimums and allow you to purchase 1 yard at a time. Some have minimums and set-up fees but the cost per yard may be lower. The typical turn-around time is 3-4 weeks, but may be more if the base fabric is out of stock.

If you're considering having your printing done in-house, digital printers typically sell for $10,000 to $70,000. Keep in mind that you'll also need to purchase equipment for curing the ink and a dedicated computer to run the printing software.

8.4   How digital textile printing improves production processes

Digital textile printing is the latest innovation in the textile industry. It helps printers to meet demands for high-quality designs, short-runs and rapid-turnarounds.

Also, its fast sampling and short-term delivery, helps customers to bring new ideas to the market faster. With digital printing, designs can be adjusted digitally on the spot. This gives digital printing companies a competitive advantage over manufacturers using traditional printing techniques.

8.4.1   Benefits of digital textile printing

High resolution, fine patterns and unlimited colour combinations.

Low initial costs as no screens have to be engraved for individual patterns.

A sustainable solution: less ink waste and high savings on energy and water.

Unlimited repeat size as it is not limited to the circumference of the rotary screen.

Digital textile printing enables printing on a large number of materials. Digital printers can be used for printing on silk fabric, polyester, viscose and numerous other organic materials like cotton, wool and linen. Our digital printing inks are most versatile, ensure a high-quality print and provide the best industry standard fastnesses.

8.4.2   Digital textile printing machines

Digital printing is driven by technology and innovation. It is presently one of the most exciting segments in the whole textile processing.

A variety of new printing systems and consumables for digital printing have been launched during ITMA 2019 in Barcelona, Spain, the world's most established trade exhibition for textile and garment manufacturing technologies.

8.4.3   Industrial single pass digital printing

The most significant development is perhaps the launch of three new industrial single pass digital printers to join MS-LaRio in the segment of high productivity ink jet printers. The new printers are Konica Minolta Nassenger SP-1, Atexco Vega One and SPGPrints Pike.

MS´s LaRio, launched in 2010, employs Kyocera printheads, a resolution of 600×600 dpi, a maximum printing width of 3.2 meters wide at speed of up to 75 linear meters per minute with 4 colours and 35 m/min with 8 colours. MS also announced during ITMA that they sold more LaRio printers to various customers including Sanitars Cotton Line (Italy), SCR (Italy), Shrijee (India) and Ruyi (China).

8.4.4   Comparison between single-pass vs multi-pass printers

1. Single-pass

A single-pass digital textile printer has no moving carriage. The printheads are mounted over the full width of the fabric on a fixed print bar, one for each colour. The fabric moves under those bars with a constant speed, so the image is built up in one stroke over the full width. This enables very high printing speeds, ranging from 20 linear meters per minute up to 40 linear meters per minute.

2. Multi-pass

In the scanning (multi-pass) digital textile printer the print heads are mounted to a carriage that moves from left to right and right to left over the width of the fabric. These printers typically reach speeds ranging from 20 linear meters per hour up to 367 linear meters per hour.

SPG Prints new single pass digital textile printer flagship model Pike isbased on Fujifilm Samba print heads. The new printer has a higher resolution of 1200×1200 dpi, a maximum printing width of 1.85 meters wide, delivering a typical productivity of 40 linear meters per minute. Pike employs Archer Technology to provide a higher printing distance (4mm) from the nozzle plate necessary for digital textile applications, enabling a wide range of fabric textures. SPGPrints announced during the ITMA exhibition that they sold the first Pike single-pass digital printer to KBC Fashion, based in Lörrach, Germany.

The Japanese company Konica Minolta announced the NASSENGER SP-1, a single-pass ultra-high speed, high productivity and high resolution digital printer. The new single pass printer system is based on newly developed print heads that can eject smaller ink droplets. It uses proprietary ink ejection control technology, which enables adjustment of ink droplet sizes, thus achieving higher-quality reproduction of fine patterns and colour gradients. For minimized downtime, the model is equipped with functionality that compensates for errors detected by print head nozzle sensors, and able to adjust density uniformly for each colour on all of the print head modules, thanks to density correction functions using image sensors.

The Chinese company Atexco (Hangzhou Honghua Digital Technology), also launched a single pass in ITMA exhibition in 201 9, the Atexco Vega One which is also based on Fujifilm Samba print heads.

8.4.5   Digital inks

Digital inks are a major cost factor in operating costs of ink jet printing on textiles. After the ink jet printers have reached productivity levels comparable or higher than screen printing, the comparably high ink cost are still a limiting factor for cost efficient digital printing. However, ink prices have already come down due to more competition and open systems. Ink margins are still high, process are far above material cost of the inks. Digital printing companies make more money with consumables such as inks, rather than with printing hardware.

The dyes contents in inks, for example in the case of reactive dyes inks, is only 10-15%, and considering that same colorants are used as in conventional printing, but in a different, more sophisticated liquid formulation, it is obvious that prices still have a long way to go down. Also, the high margins have invited more players in India and China to start an ink business which will drive down process further.

One way to stop the price erosion would be to close the open systems, to make it harder for textile printers to change the ink systems. Traditionally, Japanese companies have preferred closed systems, for example Konica Minolta covers all aspects of the printing system: the print head, printer, ink and software are all developed by them. On the other hand MS LaRio is an "open system" running with inks for various suppliers, including digital inks by SPGPrints, by the way. The cost of head replacement is a major concern for buyers in fixed-array machines. SPGPrints proposes an arrangement in conjunction with Pike® inks, in which any faulty printing heads returned would be replaced free of charge. However, this may be a show stopper for competitive inks and would reduce the bargaining power for ink buyers. A good marketing effort by SPGPrints, so to speak.

8.5 Pigment ink digital printing

In textile printing, 50% of colorants are pigments, but in digital printing the share is much less. The reason is the risk of nozzle clogging caused by pigment particles. Now, technological improvements in pigment ink formulations and printing technology, digital printing with pigment inks are getting more and more interesting. Pigment printing only require a heat-fixation step for post treatment, and therefore provides an eco-friendly process with reduction in water consumption and waste.

Some new products that were launched at ITMA 2019 included Kiian's Digistar K-Choice, a new pigment ink for the Kyocera heads, Velvet jet's new pigment printing solution for digital textile printing, Fujifilm Imaging Colorants new Pro-JET TX pigment based ink for high speed textile printing with dry heat fixation technology with allows for waterless fixation.

Also, some machines exhibited at ITMA 2019 exhibition were running on pigment ink successfully, for example the Durst Alpha 330, Colorjet Metro. Miyakoshi MTP Series, Kornit Allegro and MS – JPK Evo indicating the evolution of pigment inks to suit high speed production machines.

8.5.1 Benefits of single-pass printing

Printing speed: up to 40 linear meters per minute
Suitable for a variety of print runs on a limited amount of fabrics
Facilitates high-speed production of large quantities on the same fabric
Preferred solution for several million meters of fabric per year on a limited amount of fabrics

8.5.2 Benefits of multi-pass printing

Printing speed: up to 367 linear meters per hour

High level of flexibility in fabrics as they can be changed quickly without a high fabric loss

Suitable for gradually increasing production capacity by adding more printers step by step

Smaller footprint

8.5.3   Costs of single-pass printing

A decisive factor in choosing a printing method are the costs and how it will enable return on investment. The low initial costs, low loss of fabric during start-up and reduced use of energy and water are important benefits of digital printing. On top of that, single-pass digital printing offers you more cost-saving benefits:

Single-pass digital textile printing requires a higher initial investment

Lower service costs as there is more control over print head performance

Low ink costs

8.5.4   Costs of multi-pass printing

Digital textile printing provides you as a printer various cost-effective benefits: low initial costs, low loss of fabric during start-up, low amount of waste and reduced use of energy and water. Are you interested in multi-pass printing? The costs of this printing technique are made up of various elements:

An affordable initial investment when starting with multi-pass digital textile printing

In turn, it has the lowest depreciation costs per meters as a result of the large print capacity

The lowest ink costs as a result of the purchase of large quantities of kilos ink per year

8.6   Advantages of digital textile printing

The digital textile printing has many advantages compared to the conventional printing method. The best aspect of digital printing technology is that there is no limitation on the usage of colors or repeat size. Multiple color shades can be printed on the fabric at a time, which is not as the case of traditional printing techniques. The digital textile printing (DTP) system can supposedly produce 16 million colors and shades. Hence, the process is time saving and cost effective.

Digital printing has proved to be advantageous for designers, textile companies and retailers. It allows the user to print quickly and as little as required, and with high rate of accuracy. While in the case of traditional printing, a minimum quantity has to be printed. Thus the overall cost of producing a sample is considerably reduced. It does not even require color registration of plates or screens.

Figure 8.4

It can save the fashion designers from theft of their designs before it is released in the market. Further, it enables the designer to control and customize the design and printing process from any location. Any corrections or modifications can be made without additional cost. This promotes timely delivery, and reduces the need for inventory. In addition, digital printing can be easily done from files stored or transmitted from a computer.

Last but not the least, as pollution from textile industry has become a serious problem, digital printing technology is environment friendly and saves water compared to the conventional printing methods. It creates less pollution as there is no discharge of dyes and chemicals. However, digital textile printing system has few limitations.

These machines cannot print metallic colors. Also, with flat color printing, the machine needs to create a range of colors which it cannot accomplish. Besides, the maximum width to which the machine can print on fabric is 150cms. The rejection level of fabric printed is higher than other forms of printing.

Latest development in ink products and color management software has substantially helped to produce variety of colors with best printing quality. New technology can manage complex colors while printing. The new DTP system engages ‹Drop on demand technology›. This technology enables quick production with variety of substrates.

The current market exhibits some new innovations in DTP technologies. Some of the developments are thermal inkjet, piezoelectric inkjet, continuous inkjet, thermal transfer, electrostatic and electrophotography, each having its own unique qualities, advantages and limitations. With the latest advancement in inkjet technology, digital printing can be done on most of the fabrics. The ITMA exhibition held in Paris displayed new digital textile products presented by the original equipment manufacturers from all over the world.

Digital printing has opened new doors for textile and clothing designers in the retail market. In the future, digital printing technology will not only replace the existing printing methods, but also offer new products, opportunities and markets.

Ink jet printing is also known as digital printing. It has become the major printing technology in the desktop/network printing markets. The advent of digital color printing has opened up many new application areas for ink jet including wide-format graphic arts and increasingly industrial applications such as textiles, which, until recently, were the exclusive domain of the traditional analogue printing technologies.

As the printing industry moves towards these new industrial ink jet markets then the media, whether it be coated paper, films or textiles, becomes an integral part of the technology and knowledge of the chemistry of the interaction of the ink, colorants and the media becomes vitally important.

Ink jet is a technology wherein there is no printing master and hence only the ink drops make contact with the substrate. It is therefore classified as a non-impact printing method.

The ink jet formulation, the specific print head and the complex interactions between them have all to be considered when we start to develop total ink jet solutions for industrial applications. Ink jet is a technology wherein there is no printing master and hence only the ink drops make contact with the substrate. It is therefore classified as a non-impact printing method. (1)

8.7   Components of inkjet printer

Basically Ink jet has three basic components. These are the print head, the ink, and the medium all of which need to work well in order to produce an acceptable output.

8.8   Inkjet technologies

There are 2 types of inkjet technologies.

1. CIJ (Continuous ink jet)
2. DOD (Drop on demand).

8.8.1   Continuous inkjet technology

In CIJ, ink is squirted through nozzles at a constant speed by applying a constant pressure. The jet of ink is unstable and breaks into droplets as it leaves the nozzle the drops are left to go to the medium or deflected to a gutter for recirculation depending on the image being printed. The deflection is usually achieved by electrically charging the drops and applying an electric field to control the trajectory. The name 'continuous' originates in the fact that drops are ejected at all times.

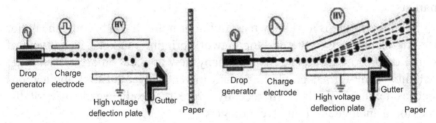

- Binary deflection
  - Uncharged droples dispended on sub strate
  - Charged droplets recirculate
- Multiple deflection
  - Uncharged droples reciruclated by gutter
  - Charged droplets deflected according to q/m ration
  - 2-dimensional writing of small areas with singe nozzle

Figure 8.5 Continuous inkjet (CIJ)

8.8.2   Drop on demand technology

In DOD ink jet, drops are ejected only when needed to form the image. The two main drop ejector mechanisms used to generate drops are piezoelectric ink jet (PIJ) and thermal ink jet (TIJ)

In PIJ, the volume of an ink chamber inside the nozzle is quickly reduced by means of a piezoelectric actuator, which squeezes the ink droplet out of the nozzle. In TIJ, an electrical heater located inside each nozzle is used to raise the temperature of the ink to the point of bubble nucleation. The explosive expansion of the vapor bubble propels the ink outside the nozzle.

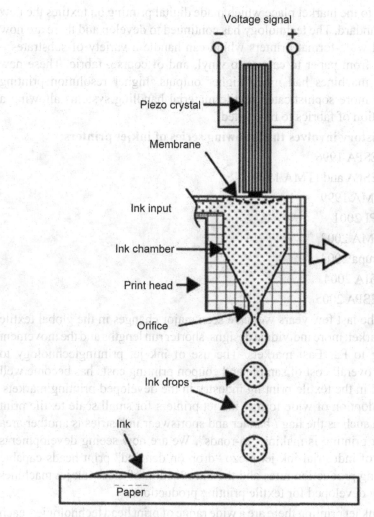

Figure 8.6 Drop on demand (DOD)

8.9   Evolution and progression of digital printing of textiles

The idea of digital printing on textiles has been around for some time. The inkjet printing technology used in digital printing was first patented in 1968. Carpet inkjet printing machines have been used since the early 1970s. Digital ink jet printing of continuous rolls of textile fabrics was shown at ITMA in 1995. In the 1990s, inkjet printers became widely available for paper printing applications. Again at ITMA in 2003, several industrial inkjet printers were introduced to the market place which made digital printing on textiles the new industry standard. The technology has continued to develop and there are now specialized wide-format printers which can handle a variety of substrates – everything from paper to canvas to vinyl, and of course, fabric. These new generation machines had much higher outputs, higher resolution printing heads, and more sophisticated textile material handling systems allowing a wide variation of fabrics to be printed.

The history involves the following series of inkjet printers;

- FESPA 1996
- FESPA and ITMA 1999
- ITMA 1999
- DPI 2001
- ITMA 2003
- Drupa 2004
- SGIA 2004
- FESPA 2005

Over the last few years we have seen major changes in the global textile printing market: more individual designs, shorter run lengths and the movement of printing to Far East markets. The use of ink jet printingtechnology to reduce the overall cost of sample and coupon printing costs has become well established in the textile printing industry in the developed printing markets. Also the adoption of wide format ink jet printers for small scale textile print production such as the flag / banner and sportswear industries is another area that ink jet printing is making "in-roads". We are now seeing developments in the use of industrial ink jet piezo "drop on demand" print heads capable of increasing production rates and the introduction of more ink jet machines specifically developed for textile printing production.

With ink jet printing there are a wide range of print head technologies, each with specific ink physical and chemical requirements that must be satisfied for the ink formulation to perform reliably in a specific printer platform. These physical and chemical specifications are very precise and require the development of textile ink jet ink formulations, which differ from normal printing pastes, in that they cannot contain the majority of chemicals, required to achieve color yield, definition or color fastness. In addition the textile ink jet ink formulations must be developed with the aim of achieving excellent operability and firing performance, together with chemical compatibility with the materials used in the manufacture of the print head.

Digital printing includes pre-treatment of fabric prior to printing process. Pre-treatment of textiles in preparation for ink-jet printing is carried out because inclusion of auxiliary chemicals and thickeners into the low viscosity ink has proved troublesome. Thus the methodology is akin to 'two-phase' conventional printing as opposed to the 'all-in' approach. In the latter case all the dyes, chemicals and thickeners required are included in the print paste, whereas in the former some of the ingredients, particularly chemicals, are applied before, or after, printing.

When printing cotton the choice has generally been between reactive dyes and pigments. The pigment printing process is simpler, as it involves three main stages (print, dry, bake/cure), whereas reactive printing has two extra processes (print, dry, steam, wash-off, dry). Pigment printing is therefore a more economical procedure but we avoid its use in inkjet printing because pigments produced much duller shades than could be achieved with dyes, and there was a tendency for nozzles to block, in other words the 'run-ability' was poor.

Reactive printing by the 'all-in› method is the normal approach for screen printing, but for jet printing it has certain dangers. As a result the jet printing of cotton, wool and silk has generally been carried out by the 'two-phase› method, the ink containing only purified dyes, the thickener and chemicals being applied to the substrate in a pre-treatment. Although the quality of the resulting prints is excellent, the extra expense of pre-treating the fabric by a pad/dry process makes the process uneconomical for anything but short runs. (4)

8.10   Reasons for pre-treatment

The main reasons for separating the dye ink from thickeners and other chemicals and applying them separately to the fabric are as follows.

- 'All-in' inks are less stable and have lower storage stability, e.g. reactive dyes are more likely to hydrolyze when alkali is present in the ink.
- Chemicals in the ink cause corrosion of jet nozzles; the deleterious effect of sodium chloride on steel surfaces is well known, for instance; inks for use in 'charged drop' continuous printers should have low electrical conductivity.
- Thickeners in the ink often do not have the desired rheological properties.
- Some chemicals can be utilized in pre-treated fabric but would cause stability problems in the ink; e.g. sodium carbonate as alkali for reactive dye fixation is acceptable on the fabric but not in the ink.

- The presence of large amounts of salts in aqueous inks reduces the solubility of the dyes; concentrated inks are required in jet printing due to the small droplet size.
- The advantage of applying thickeners and chemicals separately from the dyes is that it allows the wettability and penetration properties of the fabric to be adjusted.

8.11   Comparison between conventional screen printing and digital printing

Table 8.1 Comparison between conventional and digital printing

Comparable factors	Screen printing	Digital printing
Tools	The process involves making a stencil using a drawn/digitized image or a photograph, attaching to a screen, placing it over the desired canvas and spreading the ink over the image.	All need is a computer and a printer with ink cartridges of every color.
Efforts	Takes a lot of time consuming effort, because the screens need to be made and the process takes long	As it is so easy to operate and gives results at the touch of a key. It is relatively quicker.
Quality	Offers better quality imaging as the ink gets deeply absorbed and lasts longer. Screen printing also gives clearer edges to the image printing, because of the precision that carefully created stencils offer.	The ink does not spread because the image is directly printed on the fabric, but tends to fade quicker that the screen printed images. However, person has a colorful image to imprint, then this is an option, all the colors are present in the single image and person does not need separate screen for the same.
Cost	Costs escalate with the numbers of screens. If person want a more complex image with many colors, then individual slides for every color are created. It also required trained labor which adds to the cost. It is most apt if person want a large quantity.	The computer and printers are one time investments and digital printing is cheaper compared to screen printing as the charge is offer for per imprinted image.

8.12   Time to introduce a new product

Table 8.2 Comparison for time to introduce a new product in conventional and digital printing.

Component	Conventional	Digital
Color separation/Design editing	2 weeks	2 weeks
Digital fabric samples	–	2-4 days
Screen engraving	1 week	–
Strike off	1 week	1 week
Production yardage	3-4 weeks	1 week (low yardage)
Total	7-8 weeks	2-3 weeks

8.13   Advantages of inkjet printing

Digital printing has revolutionized the way businesses create their printed materials. It is fast, effective, and provides an alternative to the more traditional method of textile printing.

- **Quality**: When it comes to quality, nothing surpasses digital printing. Images are essentially flawless, alignment and registration issues are non-existent, and color is vibrant. Digital printers can also use the entire length of a printable item.
- **Speed:** Digital printing's ability to switch over to a new label almost instantly is another perk of using digital printing. Because there's no lost time setting up plates and printing machinery, your order is likely to reach its intended destination days, if not weeks earlier.
- **Short run printing advantage:** Digital textile printing efficiently produces designs at run lengths as low as one yard of fabric without the need for screen changes.
- **Lower water and power consumption:** Digital textile printing eliminates the substantial amount of water and electrical energy one requires for rotary screen preparation, printing and cleanup. Even greater water and power savings can be achieved with disperse/sublimation and pigment digital textile inks, which only require a heat-fixation step for post treatment.
- **Less chemical waste:** Digital textile printing results in significantly less ink usage and waste relative to screen-printing. Taking into account the additional chemistry and chemical waste from screen production, printing digitally offers a greener advantage for printing.

- **Large repeat sizes:** Digital textile printers can print large designs (e.g. cartoon characters on sheets and blankets) on roll fabric without the usual rotary screen-printing limitation in pattern repeat size.
- **Reduced production space requirements:** By not having to prepare and store customer screens for future use, the production footprint for digital printing is a fraction of the size one requires for a rotary screen print facility.
- **Less printed inventory needed:** Digital textile printing permits the option to print a design at will. This means that manufacturers with an integrated digital printing system in their production chain can keep a stock of unprinted textiles on hand to print as required. This reduces the need for pre-printed inventory of fabric that may or may not be used.
- **Sampling and production done on same printer:** By being able to print samples (strike-offs) on the same printer one uses for production, digital textile print shops can present their customers with proof samples of designs that will exactly match the final printed material.
- **Print flexibility:** Printing houses utilizing both digital and screen technologies can choose to print a small quantity of designs with different color combinations (color ways) first with their digital textile printing solutions for test the market. They can later opt to print higher volumes of the most desired color designs using rotary screen technology.
- **Variety of creative design choices for printing:** Digital textile printing provides the option to print photographic/continuous tone images, spot color pattern designs or a combination of both. This expands the creative printing alternatives for fashion and interior designers.
- **Low capital investment:** The relatively low capital investment to setup a digital textile print shop, especially compared to rotary screen-printing production, makes it possible to start small and expand as business grows.

8.14   Limitations of inkjet printing

- **Limitation of particle size:** Metallic colors cannot be printed by these machines due to large particle size.
- **Large Volumes are expensive:** Without getting too technical, digital printing presses run at a maximum of about 50 feet per minute. While

this speed is sufficient for low volume (10,000 – 15,000 item) projects, larger volume work will benefit from using traditional presses that can run at speeds between 300 and 500 feet per minute. Although traditional presses are more expensive to configure and operate, they will save you money if your jobs are very large.

- **Ink limitations:** While digital printing certainly handles color and ink well, digital inks have a tendency to fade more quickly than offset inks when exposed to direct sunlight. Also, the opacity of digital ink isn't quite up to par with offset ink, because digital ink is naturally thinner (though the difference between the two is only noticeable when dealing with clear or metallic media). There are types of laminations available to help prevent this problem from occurring.

8.15   AIMS and objectives

Fabric pre-treatment is essential for textile printing with reactive dyes to ensure efficient inkjet print performance, for example to achieve acceptable color strength and fastness properties, and to control droplet penetration and spread for optimum image quality, because the auxiliary chemicals required, such as urea, alkali and migration inhibitor, cannot normally be incorporated into the inks. Therefore, the aim of our project is:

1. To study the effect of the fabric pre-treatment on color strength and dye fixation of a digital printed cotton fabric.
2. To optimize a pre-treatment recipe and to analyze the effect of the fabric pre-treatment on color strength and dye fixation of a digital printed cotton fabric.

e chemicals which are used in the pre -treatment of cotton fabric are;

- Thickeners
    - I. Sodium alginate(Natural thickener)
    - II. Thermacol min (Synthetic thickener)
    - III. Prepajet uni (Synthetic thickener)
- Urea
- Alkali (Sodium bicarbonate)
- Anti-reduction agent (Lyoprint RG-GB)
- Penetrating agent (Lyoprint air)
- Reactive inks

8.15.1   Thickeners

Thickeners are employed in printing to preserve the sharpness of edges and outlines by countering the natural wicking effect of the substrate. In addition they hold moisture to enable dyes and chemicals to dissolve and enter the fibres during the steaming stage after printing and drying. They also modify the flow properties (rheology) of the ink or print paste. The thickening agent should not react with either the dye or other chemicals present because, if they do, an insoluble product usually results. This does not wash off and the fabric becomes stiff.

They are of different types of thickeners available in the market are as below

8.15.1.1   Sodium alginate

The chemical compound sodium alginate is the sodium salt of alginic acid. Sodium alginate is a gum, extracted from the cell walls of brown algae. A major application for sodium alginate is in reactive dye printing, as thickener for reactive dyestuffs (such as the procion cotton-reactive dyes. Alginates do not react with these dyes and wash out easily, unlike starch-based thickeners. It is poly anionic in nature. It is this property that prevents the anionic reactive dyes from reacting with the thickener, since both have negative charges and so repel each other.

The uses of alginates are based on three main properties.

- The first is their ability, when dissolved in water, to thicken the resulting solution (more technically described as their ability to increase the viscosity of aqueous solutions).
- The second is their ability to form gels; gels form when a calcium salt is added to a solution of sodium alginate in water.
- The third property of alginates is the ability to form films of sodium or calcium alginate and fibers of calcium alginates.

8.15.1.2   Thermacol min

- Chemical constitution: aqueous solution of an acrylic polymer.
- Ionic character: anionic
- Physical form: colourless liquid.
- Storage stability: store at 20°C more than 1 year.
- Compatibility: compatible with anionic and nonionic auxiliaries.

8.15.1.3   Prepajet UNI

Prepajet UNI is high concentration synthetic thickener for reactive printing with high electrolyte stability ensuring excellent colour yield & sharp definition.

8.15.2   Alkali

Reactive dyes react with cellulose under alkaline conditions to form covalent bonds between fibre and dye. There are various classes of reactive dyes, Monochlorotriazine (MCT), vinyl sulphone etc., and these require different strengths of alkali for optimum fixation. Sodium bicarbonate is generally recommended for 'all-in' pastes and inks, as it causes least hydrolysis of the dye on storage.

8.15.3   Urea

Urea is a very common constituent of print pastes as it acts as both dye solvent and hygroscopic agent (or humectant). The main functions of urea are:

- To increase the solubility of dyestuffs with low water solubility. This hygroscopic effect does not influence fixation rates significantly.
- To increase the condensate formation necessary for the migration of dyestuffs from paste to fibres.
- To form condensate with increased boiling point, thus the requirement on steam quality can be reduced.

8.15.4   Anti-reduction agent

Anti-reduction agents protect dyes against shade changes in printing, hence achieving excellent prints with good reproducibility.

8.15.5   Penetrating agent

Penetrating agent or De-aerating agents are used to remove air from print pastes and improve ink penetration in printing systems.

8.15.6   Recipe for pretreatment

Recipe -1

Ingredients	Amount (g/l) in 500ml solution
Thickener (Sodium alginate)	100
Anti-reduction agent (lyoprint RG-GB)	10
Penetrating agent (lyoprint air)	-
Urea	100
Alkali	30
Deionized water	94ml
Steaming time	5 min at 102°C

Procedure

- Add 94 ml water in beaker.
- Then add the given amount of ingredients in water in following sequence; urea, alkali, reduction inhibitor, thickener, penetrating agent.
- Stir with Stuart SS20 overhead stirrer until the preprint paste consistency becomes thick.
- Pre-treat the cotton fabric by using padding mangle at conditions of 1.4 bar pressure and 1.5 rpm speed to achieve 75-80% pick up.
- Dry the fabric at 120°C for5 min.
- Digitally print the fabric by using reactive inks.
- Steam the printed fabric at 102°Cfor 5 min.
- Fabric is then washed in the following sequence (tap water, hot water and cold water).

Recipe-2

Ingredients	Amount (g/l)
Thickener (Prepajet UNI)	80
Anti-reduction agent (Lyoprint RG-GB)	20
Penetrating agent (Lyoprint air)	6
Urea	200
Alkali	40
Deionized water	X
Steaming time	7-10 min at 102°C

Procedure:

- Add x ml of water in a beaker.
- Then add the given amounts of ingredients in water in the following sequence urea ,alkali, reduction inhibitor, thickener ,penetrating agent.
- Stir with Stuart SS20 overhead stirrer until the preprint paste consistency becomes thick.
- Pre-treat the cotton fabric by using padding mangle at conditions of 1 bar pressure and 1 rpm speed to achieve 70-80% pick up.
- Dry the fabric at 120°C for 5 min.
- Digitally print the fabric by using reactive inks.
- Steam the printed fabric at 102°C for 7-10 min.

- Fabric is then washed in the following sequence (tap water, hot water and cold water).

Recipe-3

Ingredients	Amount (g/l)
Thickener (Thermacol MIN)	100
Anti-reduction agent (Lyoprint RG-GB)	20
Penetrating agent (Lyoprint air)	10
Urea	150
Alkali	40
Deionized water	X
Steaming time	7-10 min at 102°C

Procedure:

- Add x ml of water in a beaker.
- Then add the given amounts of ingredients in water in the following sequence urea, alkali, reduction inhibitor, thickener ,penetrating agent
- Stir with Stuart SS20 overhead stirrer until the preprint paste consistency becomes thick.
- Pre-treat the cotton fabric by using padding mangle at conditions of 1 bar pressure and 1 rpm speed to achieve 70-80% pick up.
- Dry the fabric at 120°C for 5 min.
- Digitally print the fabric by using reactive inks.
- Steam the printed fabric at 102°C for 7-10 min.
- Fabric is then washed in the following sequence (tap water, hot water and cold water).

8.15.7   Post-treatment

When the pre-treated fabric has been dried and then jet printed there is usually little need to provide a drying station to dry the print, as the printing process is so slow. By the time the fabric is batched on a roll it has dried by exposure to the warm atmosphere in the room. However, in most instances fixation and washing will be necessary. This not only ensures that the full fastness properties of the dyes are realized, but also brightens and alters the colours significantly.

8.15.8   Fixation

Steaming is the process normally used to fix printed textiles. During the process steam condenses on the fabric and is absorbed by the thickener and hygroscopic agents in the printed areas. Dyes and chemicals dissolve and form extremely concentrated dye baths within the thickener film. The digitally printed fabrics were steamed at 102°C (saturated steam) in the range of 5-10 min.

8.15.9   Washing-off

After printing and steaming, washing of ink-jet printed fabric is carried out. The reason for this is that thickener, auxiliaries and loose dye should be removed under conditions where the dye is unlikely to stain white or unprinted ground shade areas.

The washing was done in three steps:
- First the fabric was cold rinsed.
- The main requirement after the first cold rinse when washing off reactive dye prints is to ensure that the temperature of the hot wash reaches a minimum of 90°C, otherwise hydrolyzed dye may not be removed. Then the fabric was washed by hot water (at the temperature of 80 - 90°C) containing 2-3 drops of surfactant (soaping SN).
- After that the fabric was again cold washed.

8.16   Digital carpet printing

Carpet decorating existed for hundreds of years, but it wasn't industrialized until the 19th century when technology became available for mass production. Today, carpet is traditionally decorated with several methods including tufting, weaving, and screen printing.

Tufting is the most widely used carpet construction method that requires dyed yarn for loop, cut, level-cut loop, and additional constructions and styles. In this process, hundreds of needles push looped yarn into a support fabric. An added backing material fixes the pile yarn and ensures the fabric's stability. Colours and patterns are applied with several methods such as piece or spinneret dyeing. While this process is the most popular for carpet construction and decoration, tufted loop pile carpet has a greater potential of snagging and running.

Weaving is another popular carpet construction method that uses dyed yarn and different weaving technologies. Most often, pile yarns are interlaced in one of several techniques that each present a different appearance. Compared to tufting, woven carpets are generally more expensive but also have a better appearance. This method also uses less pile weight.

While tufting and weaving incorporate dyed yarns for construction and decoration, screen printing is available to decorate plain carpet. This process uses flat or rotary screens in combination with pre-mixed dyes to print on low-pile constructions, mainly loop pile velour. According to experts screen printing is an outdated technology in the carpet manufacturing industry. In the 1970s, digital printing emerged in the carpet industry. At this time, Zimmer created the ChromoTronic Carpet Printing Machine that used computer controlled jets–not traditional screens–to print patterns.

Shortly after, Zimmer developed the ChromoJet, a valve-based digital press that works with pre-mixed spot colours. Up to now more than 300 installations–from two to five meters–are in operation around the world as well as many sample printers.

Today, digital inkjet carpet printers are available to create endless colours and complex patterns. Digital printing is a quickly emerging technology used for many flat surface applications.

However, this technology didn't come to be without difficulty. It was found that the main challenge was and is to use tiny droplets from inkjet print heads to obtain a good penetration on pile products. Zimmer is a pioneer and specialist in this field and developed several technologies to get maximum penetration.

While digital carpet printing expanded in use and technology, improvements are continuously made in speed and penetration on high-pile carpets as well as to minimize water usage, energy, and space.

The latest developments are underway for printing on woven polyester and wool carpet rugs. This is the perfect combination of quality and flexibility. There are three lines already in operation in the Middle East.

With a bright future ahead, digital printing is expected to revolutionize the carpet industry by decreasing stock and granting unlimited design access. Digital printing and dyeing is the future in carpet design. No limitation in colour, size, pattern, or design sampling. Furthermore, only grey material or raw fabric before undergoing dyeing or bleaching will be in stock.

The valve jet technology is a well proven technology for carpet and other high pile fabrics. The print paste is applied by nozzles instead of a screen meaning you're not bound by screen size or a long start-up process. Nearly all kinds of designs can be printed. The paste is printed on the fabric by opening and closing the nozzles.

In digital carpet printing, carpet is typically printed to directly rather than a transfer method. For a quality design, digital inkjet carpet printers require several features to handle thick and rugged materials of varying sizes.

Carpet needs a stable and robust system starting at unrolling and ending at roll-up . To do so, precise feeding throughout the line is essential for consistent quality products.

Print heads are also important for satisfactory results and to avoid challenges. When selecting a carpet printer, it is suggested manufacturers ensure the print heads are robust and can also be repaired. Additionally, carpet, mat, and rug printing requires special software to match the carpet industry's needs.

Ink types for carpet printing are selected according to the final application. For example, rugs, promotional carpets, and mats need bright colours while contract carpet and carpet tiles require enhanced light fastness properties.

Reactive, acid, disperse, and direct sublimation inks are all options when it comes to digital carpet printing. The ideal selection is acid for polyamide, wool, and silk. Disperse is ideal for polyester. Reactive works well for cellulose and cotton fabrics.

Certain substrates are typically better suited for carpet printing than others. Similar to ink, substrates should be selected based on the application. For rugs and mats, it looks a strong trend toward polyester and cationic polyester for its affordable price, softness, and recyclability. For carpet and carpet tiles, nylon is preferred due to its resilience and stability. Wool, acrylic, and cotton do not currently play a significant role.

Depending on the fibre used, carpet printing requires a post-printing process. With the Zimmer COLARIS sys-tems, nylon, wool, and cationic polyester is first washed before printing and then steamed, washed, and dried. For some applications and qualities we can skip the post-washing process.

Polyester carpets also require a pre-washing or foam impregnation. After printing, drying and high temperature fixation is used followed by reductive washing and drying again.

Dedicated digital carpet printers are few and far between—two vendors standout as pivotal players in the market.

Released in 2014, Hollanders Printing Systems offers the ColorBooster 250 Carpet Edition. The 2.5-meter wide format device is designed for carpet printing with dye-sub transfer and direct-to-textile options. It includes the proprietary Hollanders media handling system, on-board remote support system, and options for built-in humidity and temperature control.

Available by Zimmer, COLARIS is a family of high-performance, high-quality printing machines for textile and carpet materials. Devices range from 1,800 to 4,200 mili meter printing widths. The COLARIS-Carpet Printer is a digital technology working with a number of process colours that are mixed directly on the carpet surface. Available since 2016, 25 units are currently operating around the world with nearly 40 million square meters produced each year.

COLARIS works with an open ink system, which uses ink from several manufacturers and types. Therefore, we can choose inks depending on fastness requirements but logistics and prices must also be considered.

In 2018, Mohtasham Carpet became the first carpet manufacturer to install the Zimmer COLARIS solution for printing on woven polyester. This is a bold move for the company as Iran's carpet sector does not traditionally favour synthetic fibres, and so carpet production is slow on incorporating polypropylene fibres and polyester yarn.

The modified COLARIS solution is complete with a custom penetration booster system to ensure ink saturation on woven white polyester carpet–pushing ink into carpet fibers before drying, heat fixation, washing, and drying.

Ink saturation is a critical component to carpet manufacturing as it mimics the look of venerated handmade carpet. This feature is crucial to the Iranian carpet printing market. Iran has the biggest tradition in carpets. The Iranian people are very picky when it comes to penetration and the backside of the carpet, and often people buy the carpet from the backside, Lastly the Digital printing devices allow carpet manufacturers to create custom designs while cost effectively producing traditional weaves. As with any new technology, it's important for manufacturers to determine which ink types and substrates best fit the carpet's end use before selecting a digital printing device.

8.16.1   The digital textile printing market

The digital textile printing market is expected to reach USD 2.31 Billion by 2023 from USD 1.76 Billion in 2018, at a CAGR of 5.59% between 2018 and 2023.The major factors driving the growth of the digital textile printing market include the growing demand for sustainable printing; increasing demand for digital textile printing in the garment and advertising industries; shortening lifespan and faster adaptability of fashion designs; development of new technologies in the textile industry; and reduced per unit cost of printing with digital printers.The market for digital textile printing market for ink is expected to grow significantly during the forecast period. In 2017, the textile and decor and direct to garment segments contributed the major share to the digital textile printing market. With the growing demand for textiles across the globe and the rising demand for polyester fabrics in the textile industry, the textile and decor and direct to garment segments will continue to account for the maximum share of this market during the next five years as well due to the following reason

1. Growing Demand for Sustainable Printing
2. Elimination of Cloth Wastage and Easier Reproduction
3. No Waste Water from Digital Print Production
4. Energy and Ink Saving With Digital Textile Printing
5. Shortening Life-Span and Faster Adaptability of Fashion Designs
6. Reduction in Per Unit Cost of Printing With Digital Printers

9

Transfer printing

Sublimation paper printing is a technique of printing in which a negative image of the desired print is transferred to the substrate by placing the paper and then pressing with a hot plate for certain time.

Disperse dyes are used to make this kind of paper and the substrates which have an affinity to disperse dyes can be printed with this technique. Nylon and polyester have an affinity for disperse dyes. Disperse dyes have a fair to good light-fastness on dyed and printed nylon textile materials. This due to the aromatic or ring structures within the disperse dye molecules which provide the stable electron arrangement which resists the degrading effects of the sun's ultraviolet radiation.

The good wash-fastness of disperse colored nylon materials is due to hydrophobic and non-polar nature of disperse dyes making it difficult for their molecules to be washed out of the polymer system of the nylon filament or staple fiber.

Sublimation printing transfer mainly transfer onto polyester, or nature fiber less than 30% polymixed fabrics.

Sublimation paper, Sublimation ink, Fabric is the main factors influence the result.

Firstly, sublimation paper, sublimation paper need be coated, if ok, you can use double coating sublimation paper.

Sticky sublimation paper is a good choice, because the nylon fabric is very smooth, will slide when make heat transfer easily. Sticky sublimation paper can stick with the fabric, fix to anti ghosting.

Secondly, Sublimation ink is an important factor will influence the whole sublimation printing. Nylon is a specially polyester fabric, it's very smooth, the dye can't be absorded into the fiber easily. Choose good brand sublimation ink is a better way.

9.1   Disperse dyes for printing on polyester

The right combination of fabric and dye is the first step on the road to a beautiful and durable print. We have specialised in printing on polyester and polyester blends, and we always recommend using the transfer technique for these fabrics.

Heat transfer printing is a two-step printing technique, which involves a dye sublimation process. We work with disperse dye inks which are the only colorant with the ability to sublime.

9.1.1   The sublimation process

The ink printed on the transfer paper turns from a solid into a gas without becoming liquid at any time. The process takes place in the calender at high temperatures.

At the carefully set temperature, the molecules of the polyester fibres open. You may say that the fibres will soften and let the ink (turned into gas) come through.

When removing the polyester fibres from heat after the sublimation, the molecules of the polyester fibres will settle again and absorb the ink completely. This fusion between the fabric fibres and ink makes this technique incredibly reliable. The colour is absorbed in the fabric instead of sitting on the surface of the fabric like other textile printing techniques.

The colour fastness is intact and there is no need for further finishing or washing of your fabric subsequent to the sublimation process. Some specially treated textiles, i.e. Trevira CS, however, will become stiff in the calender and do need washing to obtain their original feel.

9.2   What are disperse dyes?

*   Disperse dyes are insoluble in water after print or have a very low solubility in water
*   Disperse dyes are used for dyeing of synthetic fibres, especially polyester
*   Disperse dyes have a good light fastness with an assessment of 4-7
*   The washing fastness is good with an assessment of 4/5-5
*   The colours undergo no chemical modification during printing

9.2.1   Ghosting – shades on the fabric after printing

Ghosting is the term used for describing a blurred pattern or unwanted shades on the fabric after printing. The most common cause of ghosting is the transfer paper quality. If the transfer paper is very thin, as the paper used for conventional textile printing, the ink risks penetrating the paper and causing trouble. The transfer paper in our digital printers is thicker and denser, and we do not have any problems of ghosting.

Hasty rolling may also cause ghosting on the fabric. It is also important to leave time for adequately cooling of the fabric before rolling. Our equipment and procedures carefully allow for space and time for cool-down and quality control subsequent to the sublimation process in the calender.

We have not had any incidents of ghosting in our digital print production.

9.2.2   Bleeding – colours "bleeding" on the fabric

Bleeding does not occur when printing on pure polyester.

In principle, bleeding may occur in all types of polyester blends, but we mostly encounter the challenge in some polyester blends with elastane (lycra).

We can easily print on a polyester fabric blended with elastane. We just need to take certain measures to ensure a satisfactory result. Most often, the elastane lies on one side of the fabric and in that case, we will of course make sure to print on the other side.

The elasticity of elastane (lycra) comes from softeners. During the print process, the grease of the softener may cause the colours to "bleed". Normally, a thin protective polyester thread is spun around the fine lycra thread. However, it may be difficult to wash out the added grease between the threads and therefore, bleeding may occur.

We always test new fabric to learn how the ink reacts to the fabric in question.

9.2.3   The digital transfer printing technique reduces the consumption of dye

Digital transfer printing has even more side benefits. With the digital technology, we are able to control the dyes 100%. This gives us great design advantages, and we use no more dye than necessary. The polyester fibres fully absorb the thin layer of ink on the transfer in the sublimations process.

The colours will not set off or in any other way come off during usage. The exceptional durability property is one of the reasons why digital printed polyester textiles are popular within medical textiles and other textiles that require special hygiene and healthcare considerations.

Apart from pottery, the technique was used on metal, and enamelled metal, and sometimes on wood and textiles. It remains used today, although mostly superseded by lithography. In the 19th century methods of transfer printing in colour were developed.

Transfer printing could be supplemented with colour added by hand, or gilding, and this technique was used from early on. The use of multiple transfers, each with a different colour, was introduced quite early when different areas were printed in each colour, for example, a plate with the centre in one colour, and the border in another. It was more difficult to build up a full polychrome image, but this was perfected by Messrs F&R Pratt of Fenton in the 1840s.

Experimented with stencils, and some pieces mix these techniques. About 50 pieces are known to survive.

Printing on enamel probably began around 1753 (a letter of Horace Walpole dated 7 September 1755 mentions a printed Battersea box), and by around 1756 his process was being used on some Bow porcelain, although the results were not excellent, perhaps as the glaze was «too soft and fusible», giving a tendency to blur the image. The colours of the 1750s were a «purplish or brownish black» or a «beautiful warm brick-red». By around 1760 there was some underglaze printing in blue.

Five years after Brooks›s first patent attempt, in 1756, John Sadler (in partnership with Guy Green) claimed in a patent affidavit that they had spent the past seven years perfecting a process for printing on tiles and that they could "print upwards of Twelve hundred Earthen Ware Tiles of different patterns " within a period of 6 hours. Sadler and Green printed in Liverpool, where their trade included overglaze printing on tin-glazed earthenware, porcelain, and creamware.

Transfer printing on porcelain at the Worcester porcelain factory in the 1750s is usually associated with Robert Hancock, an etcher and engraver, who signed some pieces and had also worked for Bow. Richard and Josiah Holdship, the managers of Worcester, were very supportive and involved with Hancock's work. By the mid-1750s the Worcester factory was producing both underglaze prints in blue and overglaze prints, predominately in black. Some printed pieces were in complicated shapes and included gilding, showing that the technique was at this point regarded as suitable for luxury products.

From 1842 the United Kingdom Patent Office introduced a system of registered marks, usually impressed or printed on the underside of pieces. Transfer-printed designs were easily registered by submitting the transfers printed on paper. The technology of transfer printing spread to Asia as well. Kawana ware in Japan developed in the late Edo period and was a type of blue-and-white porcelain.

Transfer printing is the term used to describe textile and related printing processes in which the design is first printed on to a flexible nontextile substrate and later transferred by a separate process to a textile

Transfer Printing was first developed to embellish ceramics, not clothing. The technique was born around the 1750's in England and quickly spread to other parts of Europe where it caught on. Back then the process involved a copper or steel plate or roller that was engraved with a decorative element. The roller or plate would have its surface covered with ink and later would be pressed or rolled over the desired piece. It wasn't an easy procedure (or fun) by any means, but it was still quicker than hand painting and the result was similar enough.

Thermal Transfer Printing, which is the kind of technique mostly used today, didn't come until much later. It was invented by a corporation called SATO during the late 1940's in the US

9.3   What is transfer printing?

The idea of transferring a coloured object from one material to another has been in use for thousands of years. Ongoing development and patenting of the different techniques have been going on forever. In modern time, the heat transfer print is the technology used the most. In the heat transfer print process, the colorants sublime from the transfer paper to the synthetic fibre material under high temperatures.

During the sublimation, the colorant turns from a solid into a gas without becoming liquid at any time. It evaporates into the textile fibres. The molecules of the synthetic fibres open at the high temperatures; the fibres soften and absorb the ink. Once removed from the heat, the fibres will cool down and settle around the colorants. This technique provides very high colour fastness.

Only disperse dyes have the ability to sublime. The size of the molecules determine whether the colorant is suitable for transfer printing.

9.3.1   Transfer printing has many advantages

- Multiple vibrant colours
- Sharp details
- Sorting of misprints on paper instead of fabric
- Quick change from one design to the next
- Transfer of ink and settling take place simultaneously – no need for further finishing

9.3.2   Transfer printing versus direct printing on textiles

scan htp recommends transfer printing on polyester to obtain the best result of accurate vibrant colours and sharp graphic details.

9.3.3   Transfer printing ensures unique rendering of your design

We have specialised in printing on polyester and polyester blends. We use the transfer technique and therefore, print the pattern on specially coated transfer paper. With digital transfer printing, our choice of colours is unlimited. The transfer from paper to fabric takes place in a separate process in the calender. The ink sits on the transfer paper in a very thin layer. In high temperatures in the calender, the ink will want to escape the paper and will be absorbed by the fabric (sublimation).

The transfer technique translates colours and detailed graphic lines perfectly.

9.3.4   Direct printing gains foothold, but this technology still has its limits

Direct printing with disperse colours is a quite new technology which gets attention in the textile industry. However, so far the technology involves many precautions and of course extensive finishing of the fabric to obtain a durable result.

Disperse colours will sit on the fabric in a thin layer and would be difficult to control if printed directly on the fabric. Heat causes the ink to sublime, and if you choose to print directly on fabric, you must include a subsequent heat treatment to settle the colorants. However, it will be difficult to obtain an even heat treatment and thus avoid light differences on the fabric.

9.3.5   Transfer prints are ready to use

Transfer prints usually do not require any further finishing. They are ready for immediate use. We do work with certain exceptions of specially treated fabrics like Trevira CS, which will stiffen from the sublimation process. The fabric gets its original feel back once we wash it.

9.3.6   Experiment on paper

The optimum printing technique is always a matter of design and demand for colours and details. The transfer technique allows you to experiment on the transfer paper, whereas printing directly risks wasting expensive fabric, and you may have to start all over again. Read more: Which printing technique may perfect your project?

Transfer printing is the term used to describe textile and related printing processes in which the design is first printed on to a flexible nontextile substrate and later transferred by a separate process to a textile. It may be asked why this devious route should be chosen instead of directly printing the fabric. The reasons are largely commercial but, on occasion, technical as well and are based on the following considerations.

1. Designs may be printed and stored on a relatively cheap and nonbulky substrate such as paper, and printed on to the more expensive textile with rapid response to sales demand.

2. The production of short-run repeat orders is much easier by transfer processes than it is by direct printing.

3. The design may be applied to the textile with relatively low skill input and low reject rates.

4. Stock volume and storage costs are lower when designs are held on paper rather than on printed textiles.

5. Certain designs and effects can be produced only by the use of transfers (particularly on garments or garment panels).

6. Many complex designs can be produced more easily and accurately on paper than on textiles.

7. Most transfer-printing processes enable textile printing to be carried out using simple, relatively inexpensive equipment with modest space requirements, without effluent production or any need for washing-off.

Against these advantages may be set the relative lack of flexibility inherent in transfer printing: no single transfer-printing method is universally applicable to a wide range of textile fibres. While a printer with a conventional rotary-screen printing set-up can proceed to print cotton, polyester, blends and so forth without doing a great deal beyond changing the printing ink used, the transfer printer hoping to have the same flexibility would need to have available a range of equipment suited to the variety of systems that have to be used for different dyes and substrates using transfer technology.

In addition factors such as stock costs, response time and so on do not always apply and unlike dyers, most printers are able to operate without steaming or washing by using pigment-printing methods. Thus a balance exists which not only permits but even requires the coexistence of direct and transfer printing. The relative importance of the two methods consequently varies with fluctuations of the market, fashion and fibre preference.

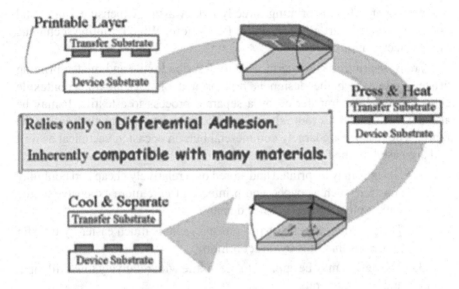

Figure 9.1 Transfer Printing Process

A great many methods of producing textile transfer prints have been described in the literature. Many of them exist only in patent specifications but several have been developed to production potential. They may be summarised most conveniently as below.

Transfer Printing has its advantages and its disadvantages.

On the good side: It's fairly simple (you don't really need a master's degree), the equipment is relatively inexpensive, especially when compared to DTG printers, and it can reproduce high quality, complex images. It is also one of the best techniques to use for full-colour prints.

On the bad side: It is slower than other procedures (still faster than hand painting, though), it's got limitations onto which types of fabrics it can be printed; those sensitive to high temperatures are a no-no, and there might be some restrictions on the reproduction of darker shades.

It may be asked why this devious route should be chosen instead of directly printing the fabric. The reasons are largely commercial but, on occasion, technical as well and are based on the following considerations.

1. Designs may be printed and stored on a relatively cheap and nonbulky substrate such as paper, and printed on to the more expensive textile with rapid response to sales demand.

2. The production of short-run repeat orders is much easier by transfer processes than it is by direct printing.

3. The design may be applied to the textile with relatively low skill input and low reject rates.

4. Stock volume and storage costs are lower when designs are held on paper rather than on printed textiles.

5. Certain designs and effects can be produced only by the use of transfers (particularly on garments or garment panels).

6. Many complex designs can be produced more easily and accurately on paper than on textiles.

7. Most transfer-printing processes enable textile printing to be carried out using simple, relatively inexpensive equipment with modest space requirements, without effluent production or any need for washing-off.

Against these advantages may be set the relative lack of flexibility inherent in transfer printing: no single transfer-printing method is universally applicable to a wide range of textile fibres. While a printer with a conventional rotary-screen printing set-up can proceed to print cotton, polyester, blends and so forth without doing a great deal beyond changing the printing ink used, the transfer printer hoping to have the same flexibility would need to have available a range of equipment suited to the variety of systems that have to be used for different dyes and substrates using transfer technology.

In addition factors such as stock costs, response time and so on do not always apply and unlike dyers, most printers are able to operate without steaming or washing by using pigment-printing methods. Thus a balance exists which not only permits but even requires the coexistence of direct and transfer printing. The relative importance of the two methods consequently varies with fluctuations of the market, fashion and fibre preference.

A great many methods of producing textile transfer prints have been described in the literature. Many of them exist only in patent specifications but several have been developed to production potential. They may be summarised most conveniently as below.

9.4   Sublimation transfer

This method depends on the use of a volatile dye in the printed design. When the paper is heated the dye is preferentially adsorbed from the vapour phase by the textile material with which the heated paper is held in contact. This is commercially the most important of the transfer-printing methods.

9.4.1   Melt transfer

This method has been used since the 19th century to transfer embroidery designs to fabric. The design is printed on paper using a waxy ink, and a hot iron applied to its reverse face presses the paper against the fabric. The ink melts on to the fabric in contact with it. This was the basis of the first commercially successful transfer process, known as Star printing, developed in Italy in the late 1940s. It is used in the so-called 'hot-split' transfer papers extensively used today in garment decoration.

9.4.2   Film release

This method is similar to melt transfer with the difference that the design is held in an ink layer which is transferred completely to the textile from a release paper using heat and pressure. Adhesion forces are developed between the film and the textile which are stronger than those between the film and the paper. The method has been developed for the printing of both continuous web and garment panel units, but is used almost exclusively for the latter purpose. In commercial importance it is comparable with sublimation transfer printing.

9.4.3   Wet transfer

Water-soluble dyes are incorporated into a printing ink which is used to produce a design on paper. The design is transferred to a moistened textile using carefully regulated contact pressure. The dye transfers by diffusion through the aqueous medium. The method is not used to any significant extent at the present time.

9.5   What's next for transfer printing?

We're never quite sure about the technology of these techniques, but after seeing this video, everything might be possible. Some savvy people in Barcelona discovered a way to transfer prints using water.

9.5.1   Transfer printing: Properly explained

So to wrap this up let's simply explain how Transfer Printing works. Let's take a look at the different stages of transfer printing:

9.5.2   Image selection

You can choose virtually any image for transfer printing. It's way of transferring allows for both complex (with many colours) and simple (with a few colours) images to be printed. Thanks to the fact that the design is printed onto paper first instead of the garment, it allows for more intricate details to show when compared to DTG that may have a more "blurry" finish. As with any other printing technique, it is advised that the original file be a high-quality one (300 dpi) to ensure the best possible result.

9.5.3   Heat transfer

Once the design has been selected, it is printed on a special heat transfer paper (I told you there was magic involved) which is then positioned on the garment. The (magical) paper is later squashed against the fabric using a heat press. It is left this way for the amount of time is necessary for the heat to do its job. After the required amount of time has passed, the press if lifted and the garment is left alone to cool down. If everything went well, then you should have a quality finish t-shirt.

9.6   How does transfer printing it work?

Most professional T-shirt printers nowadays use a more sophisticated version of the simple iron-on method, but the basics are still the same. What happens is that the heat transfer machine releases the right amount of pressure, holds the garment in place and has a consistent temperature which allows the colour pigments to be transferred from one surface to the other. Heat transfer literally melts the image onto the fabric.

9.6.1   Heat transfer paper

It's advisable to use commercial heat transfer paper as this will give the image a much better quality finish, lasts longer and won't fade, bleed or peel. Cheap paper is not suitable for professional looking print since it is likely to show a line around where it's cut and have that awful shiny finish, making the garment look very 'homemade'.

Benefits
- Inexpensive
- Good for small quantities
- Can print complex images with many colours and intricate designs
- Prints on any garment regardless of colour
- Easy for amateurs
- Clean (screen printing can be very messy)

Disadvantages

- Not practical for large quantities
- Not as flexible when it comes to printing on different kinds of materials
- Each design must be cut one by one

9.6.2   Heat transfer printing for small businesses

Heat transfer machines are relatively cheap, easy to use, lightweight and don't take up much space. This kind of printing can be done on demand, eliminating the need for holding stocks or large print runs. Just print when people place orders, as opposed to printing, keeping garments in stock and hoping you'll receive hundreds of orders. This can be beneficial for start-ups offering small quantities of specially designed t-shirts as there are virtually no extra costs involved.

Sublimation Printing: Sublimation is a transfer printing process. It is pigment based printing. Any type of design having greater complexity can easily be developed on the fabric surface .The sublimation transfer printing process consists of dye transfer in the vapour form from the paper to the fabric and this process is used on the commercial scale for transfer printing of 100% polyester fabric . In sublimation printing machine, image is first engraved on a copper plate and then pigment is applied on this plate. The image is then transferred to a piece of paper . Here, the printed design is transferred from paper to fabric surface through heat pressing and then delivered after inspection. Sequence of sublimation printing process:

Design creation and input in printing machine

↓

Color selection

↓

Printed paper output

↓

Check to match with approved standard

↓

Change dyes if required otherwise proceed

↓

Design transfer on fabric through heat press (Normally for 30 to 40 second in 200 °C)

↓

Inspection

↓

Delivery.

9.7   What is cool transfer printing of textiles ?

For the first time in 1970 in order to overcome the disadvantages of screen printing industry, the idea of combining methods of paper and fabric printing were presented and transfer printing was born. Transfer printing is a type of printing in which a color scheme of a printed paper will transfer on a cloth or polymeric film.

Modern textile printing incorporates a wide variety of colorants and technologies to produce a diverse array of printed textile products at various stages in the product development process. Today, rotary screen printing accounts for the majority of cotton printed clothing; however, newer printing technologies such as digital printing and cool transfer printing are attracting attention because of their reduced environmental footprint, and seemingly limitless color and design capabilities. Traditional transfer printing applies a print pattern from one substrate, typically paper, to synthetic fabrics by applying heat and pressure. Cool transfer printing is a newer process that transfers a design from paper to fabric without the use of heat. Due to high market interest, cool transfer printing technology has evolved to achieve high quality digital printing on cotton fabrics. Water-based cool transfer printing technology has developed very fast over the last two years in China for both natural and synthetic fiber. It has high production yield, low production cost as conventional rotary and flat screen printing. Printing of super thin polyamide woven and knitting fabric (8D, 15D, 20D) have encountered many problems by conventional rotary and flat screen printing technique as well as digital inkjet printing.

Cool Transfer Printing is a process that transfers a print design from paper to fabric. A special coated paper is printed with the desired pattern which is then transferred to the fabric under ambient conditions.

Market interest in transfer printing on cotton fabrics has been strong for many years; however, standard heat transfer printing was achievable only for polyester and nylon fabrics, and the transfer prints available for cotton offered poor hand and were limited in terms of inks available. Today, cool transfer printing technology makes it possible to transfer digital quality prints to a cotton fabric under ambient conditions.

Cool transfer printing provides the flexibility to print on a variety of fabrics, the photo-realistic print quality of digital printing, and production rates comparable to rotary screen printing. Even with the over printing technique, cool transfer printed products maintain the soft, natural feel of the cotton fiber. A recent innovation to the cool transfer printing technology is the capability to print on both sides of the fabric during one pass on a duplex printing machine.

The technology allows for easy diversification of pattern design and coloring, and achievement of multiple print/dye effects on single fabric. Although dye selection is the ultimate determinant of fabric colorfastness, in testing, cool transfer printing has exhibited better colorfastness compared to digital printed textiles, due to better penetration of the dye into the fiber.

Key Features
- Transfer print design from paper to fabric
- No heat required
- Typical speeds: 20 to 40 yards per minute
- Best for: High quality cotton fabrics, photo-realistic images, brilliant colors, soft and drapeable fabrics.

Advantages
- Ability to produce photo-realistic images
- Comparable speeds to rotary screen printing
- Maintains soft, drapeable feel
- Increased image brightness
- Reduced water, dye and energy consumption, less waste
- High dye transfer and colorfastness

Disadvantages
- Newest technology
- More expensive than rotary printing, but typically less expensive than digital printing on per yard basis
- No raised surface special effects, e.g., glitter, mother of pearl, metallic flakes, puff

9.7.1   Environmental advantage

Fewer process steps combined with lower consumption and output of WECs (water, energy and chemicals) make cool transfer printing a compelling environmentally friendly alternative to other cotton print methods. A key difference between traditional transfer printing and cool transfer printing is the elimination of heat in the process. Cool transfer printing is executed at room temperature, therefore not requiring heat. Unlike other methods of printing, up to 90% dye transfer is achievable with cool transfer printing. With greater dye utilization, there is less dye lost in the printing process, and therefore no need for extensive removal of unfixed dye. The reduction of steps in the printing process provides both a cost effective and environmentally friendly alternative to digital and rotary screen printing due to minimized water usage. Cool transfer print rolls can also be used multiple times, and used printed transfer paper can be repurposed for other applications such as packaging.

Although cool transfer printing offers many advantages for printed cotton fabrics, the technology is relatively new. Therefore the cost and performance are key barriers to adoption into widespread production. While cool transfer printing cannot replace the special effect printing possible with rotary screen printing (e.g., glitter or puff), its speed, limited WEC footprint, strong color depth and colorfastness, and image quality make it a strong competitor.

Also called COOLTRANS Technology is a green and eco-friendly production technology. It has following advantages

1. *Increases the added-value* of textile printing: finer lines, sharper contour, smoother gradation of prints.

2. *Reduces water* and *energy consumptions, less waste effluents* and *pollution:* less dye consumption, higher dye fixation, cool transfer temperature, less soaping water discharge.

3. *The printing effect is higher than direct textile inkjet printing:* higher color yields and more delicate printings on the transfer paper/file than the direct printings on the textile

9.8   Comparison of printing mechanisms

Parameters	Cool trans	Heat transfer
Process	Transfer: High pressure, room temp. Fixation: High temp, low pressure.	High temp. & high pressure transfer at the same time.
Dyes	Suitable for reactive, acid, disperse and other ionic dyes.	Only for disperse dyes.
Solution system	All water-based system. Eco-friendly.	Majority alcohol based inks. Harmful VOCs and not eco-friendly.
Penetration	High penetration & adjustable.	Low penetration. Ring dyeing.
Techniques	Inkjet, gravure printing.	Inkjet, gravure printing.
Process	High transfer rate and residual inks are easily soluble in water. Good recyclability.	Low transfer rate and residual inks are not easily soluble in water. Poor recyclability.

Garment printing

We all know about Garment Dyeing but we should also know about Garment Printing.

There are quite a few options but the two most popular are

1. Direct-to-garment (also known as DTG)
2. Screen printing .
3. Dye Sublimation.
4. Heat Press Printing:
5. We should discuss the systems how they work.

10.1   Direct-to-garment printing (DTG)

It is a printing method that sprays the ink onto the garment using inject technology. The inks then soak into the fibers of the garment. It's sort of like printing on paper, except on clothing.There are DTG printers which offer extensive colour options which means you can print detailed designs and photorealistic images with virtually no colour limitations. This can be important for those businesses that want to experiment with colour and design. It is a hassle-free way to get the products ready for your customers.

Most third-party printers have no order minimums for DTG products, so you don't have to worry about keeping stock. This printing method also enables businesses to use printing services on-demand.*With print-on-demand, your designs are printed on products only when orders come in. This means you don't need to buy large quantities of products in advance.*

With DTG, the entire design is printed in one go (unless you print colour on dark fabrics—that requires a white underbase layer). And since DTG doesn't have a color count, there's no extra setup time to start printing, so your order's fulfilled as soon as the print file comes through.

This printing model gives you the freedom to introduce new designs or enter new markets without losing any money. If the product doesn't sell, you can discontinue or replace it.

Like the screen printing method, it is also popular among hobbyists. It easy to do but it will require knowledge and skills if you are not going to get professional help. All you need for it work is the textile printer and ink .

DTG works like a paper printer in the office. The only difference is that, in this case the ink is going to the fabric. You first need to upload the design to the computer, which then creates uniqueness and creativity.

The Direct To Garment method is excellent for printing out super-complex design. It gives a soft feel on the hands when you touch the final design. The ink used doesn't create thick extra layers on the shirt, because it is thin.

However it has one problem. You can end up with a low resolution design with dot patterns if you have an inferior textile printer. You can›t use this method to create design for dark fabric, because the ink is thin and won›t look well.

DTG is least durable and the fabric may fade in a year or less.

Often referred to as digital garment printing. It is a process of printing reactive ink onto textiles and garments using specialized or modified inkjet technology which penetrates the fabric, unlike screen-printing which sits on top of the cloth.

This printing method is perfect for long lengths of fabric as a repeat image is created through the Adobe suite and sent to the printer as a file.

The inks, unlike those used for screen-printing, are very thin and do not contain a fixative. The fabrics that are used for DTG are pre-treated with a fixative.

These fabrics include silk, cotton, rayon, wool and linen as well as mixes of the aforementioned. Inkjet printing on fabric is also possible with an inkjet printer by using fabric sheets with a removable paper backing.

Today, major inkjet technology manufacturers can offer specialized products designed for direct printing on textiles, not only for sampling but also for bulk production. Printing onto nylon and silk can be done by using acid ink.

Reactive ink is used for cellulose-based fibres such as cotton and linen. Inkjet technology in digital textile printing allows for single pieces, mid-run production and even long-run alternatives to screen printed fabric.

Advantages:
- There are many companies that specialize in digital printing. Local companies such as The Silk Bureau offer a wide range of pre-treated fabrics such as silk and silk mixes, linens and cotton, stretch fabrics, light and heavy polyesters and rayon, viscose and modal.
- The process is very fast.

Disadvantages*:*

- There can be, on occasion, a loss of some fine-image detail by most commercial digital printing processes.
- The reactive dyes used are more difficult to process.
- After printing, the fabrics require hot steam fixation and extensive washing. This leads to more time spent and the use of more equipment in the production process.
- All of these processes result in an expensive product.

Costs:

- Digital printing is generally expensive if applied to silk and wool.
- Cotton and linens are slightly cheaper. It's a good idea to test areas of your artwork prior to printing full-lengths, just to ensure satisfaction with colour resolution and density of artwork.
- Some companies do offer a test service on 0.5 metres of fabric which reduces your printing costs in the long term should there be any mistakes that need resolving.
- As you can see, the choices are many. Depending on your budget, need and volume – printing is accessible to everybody.
- Advances of technology have really created numerous opportunities in this area and as a result, fashion is vibrant and diverse.
- The only limit one can encounter where print is concerned is imagination.

10.2   Screen printing

This printing technique pushes the ink through a woven mesh stencil onto fabric. The ink doesn't soak into the fabric, it lays on top. Back in the day, screen printing was the only way companies could print custom products, like t-shirts with a brand logo, in bulk.

With screen printing, a special screen has to be made for each element of your design. Once that is done, each color of the design is applied layer by layer onto the garment. So, the more layers your design has, the longer it will take to print it.It is mostly used for simple designs in fewer colors. But that doesn't mean you can't create outstanding prints using typography, line drawings, shapes, and symbols.

Screen printing can be cost-effective for orders with simple graphics (ideally with one solid colour), but it's unsustainable in the long run, especially if you want to offer a variety of colourful designs.

Why? Because it requires upfront investment and quite a bit of your time to get the designs print-ready.

Most third-party print services have order minimums that can be anything from 5 to 100 items. The bigger the quota is, the more likely you'll have to worry about where to store the printed garments if you don't want to have piles of clothing laying around your home or office. And this can be financially challenging if you're just starting out or want to experiment with new designs.

The final price of your order is also influenced by the number of colors in your designs and the screens that are needed to print it. All of which you'll have to discuss directly with the printer.

Some printers will ask you to submit print files separates into layers of each color. They might also ask you to specify the color codes for inks they should use to print your design. So if you're looking for an *upload your design and forget about the rest* type of experience, screen printing probably isn't the right choice for you.

It is the most common with the pros. Another name for this design is silkscreen printing.

For this method to work, you will need a stencil, and a nylon mesh. You set up the stencil, which is held by the nylon mesh. A water proof material will block the negative space that you want to design.

The negative space is the only part permeable for the dye. Ink is then flooded into the screen.

The screen printing method is ideal for printing high-quality t-shirt designs, because it guarantees ultimate replication of the design. The screen printing methods gives a shirt a unique, cute and impressive look that customer love.

Most companies that use the screen printing methods for mass production of t-shirt designs for your business. If you want a single or unique design for yourself, it may be impossible.

This method is suited for one color per screen. It is not ideal for super complex multi-color designs.

10.3   What's better: DTG or screen printing?

DTG and screen printing yield fine quality prints, but they differ in method and cost. DTG uses a printer to spray the ink into a garment, while screen printing layers the ink on top of the fabric. Most importantly, DTG enables order fulfillment on-demand with no upfront cost, when screen printed shirts can only be ordered in bulk.

If you still can't decide which printing method is right for you, use this comparison table:

	Screen printing	DTG
High-quality prints	Yes	Yes
Detailed designs	No	Yes
Unlimited color palette	No	Yes
Order minimums	Yes	No
On-demand fulfilment	No	Yes
Good for bulk prints	Yes	No
Requires upfront investment	Yes	No

Whether you want to scale your business using DTG or screen printing is completely up to you. Before you make your final decision, consider:

- the complexity of your designs,
- your product quantity needs,
- your willingness to invest in the stock upfront.

If you're looking for the easy and effective way to grow your business, DTG on-demand is your best bet. With this printing method, it's easy to introduce new designs to your store risk-free, and you don't have to worry about keeping stock and shipping the items yourself. All that gives you more time to experiment with your product offering and marketing.

10.4   Dye sublimation

Dye sublimation works well on light shirts or fabrics. It cost to produce a dye sub but in the end, the customer ends up with a professional done shirt by experts.

To bring out the best results in this method, you need to have deep knowledge in dye sublimation. Dye sublimation is perfect for printing polyester like cream fabrics. They are unforgiving when you bring cotton fabric to the scene.

You will need a special type of dye which is liquid. The liquid dries up when it meets the polyester fabric. When it has dried well, it solidifies on the fabric, then you can apply heat and pressure for sublimation to take place.

When the solid fabric is exposed to heat and pressure it turns into gas. The fabric molecules then expand and the gas sips between the gap that appear in the cloth. In the end, the molecules contracts again when you remove the heat.

Dye sublimation is excellent for printing polyester shirts with minimal enduring designs. Shirt done through sublimation are durable and looks cute.

Dye-sublimation printing is a digital printing technology (part of heat press) using full-colour artwork that works with polyester and polymer-coated fabrics.

Also referred to as digital sublimation, the process is commonly used for decorating apparel, and other items with sublimation-friendly surfaces.

The process uses the science of sublimation, in which heat and pressure are applied to a solid, turning it into a gas through an endothermic reaction without passing through the liquid phase.

In sublimation printing, unique dyes are transferred to sheets of "transfer" paper, via liquid gel ink, through a piezoelectric print head.

The ink is deposited on these high-release inkjet papers, which are used for the next step of the sublimation printing process.

After the digital design is printed onto sublimation transfer sheets, it is placed on a heat press along with the material to be sublimated.

In order to transfer the image from the paper to the material, it requires a heat press process that is a combination of time, temperature and pressure. The heat press applies this special combination, which can change depending on the material used, to "transfer" the sublimation dyes at the molecular level into the material.

The end result of the sublimation process is a nearly permanent, high resolution, full colour print. Because the dyes are infused into the material at the molecular level, rather than applied at a topical level (such as with screen printing and direct to garment printing), the prints will not crack, fade or peel from the material under normal conditions.

Advantages:
- Images are permanent and do not peel or fade.
- Dye does not build up on the fabric.
- Colours can be extraordinarily brilliant due to the bonding of the dye to the transparent fibres of the synthetic fabric.
- Truly continuous tones can be achieved that are equivalent to photographs, without the use of special techniques such as half-screen printing.
- The image can be printed all over the entire item, with no difficulty in printing all the way to the edges.

Disadvantages:
- The printer speed is low.
- Any creases in the apparel during printing leave blank spots behind.
- **Costs:** These are dependent on the size and type of material selected.

10.5   Heat press printing

If you have small orders of shirt, it is economical to use the heat press printing. It saves you resources and time.

This method incorporates a design printed on a special paper called transfer paper. Here is how it works.

You take a shirt, press it on the best heat transfer vinyl, then apply heat and pressure. A more similar approach like the dye sublimation with slight differences. You do the process until the heat softens the dye on the paper until it gets a beneath the cloth.

After that you take the glossy paper transfer and strip off the dye which leaves you with the intended design on the shirt.

If you are working with super complex designs, then heat and press method is ideal for you. However, it won't work well with dark fabrics because of its translucent dye.

Shirts printed through heat press printing are durable.

As the name suggests, this printing technique uses heat and pressure to inscribe the design on the garment. An average heat press machine consists of a hot plate, temperature and pressure controller, a clam-like handle to press down on, a timer and a soft rubber pad onto which a t-shirt or a garment goes.

You set the temperature to 180-190 degree Celsius, place the garment on the rubber pad, design/print your pattern that you want to print, on a transfer paper. Place the paper face-down and press the clam handle for 13 seconds (or until the timer runs off).

Lift the handle, stretch the garment so that you can peel off the transfer paper easily. You will see that the design has been printed on the garment now.

If you want to do the same at the back, just put silicone paper on the design you've just printed (for its safety and to avoid cracks) and print on the backside as you did above.

Remove the transfer paper from the back and silicone paper from the front. Stretch the fabric a little and you're good to go.

Advantages: Easy and quick to do. It allows a variety of colours in one go.

Disadvantages: Good for smaller businesses/quantities as this is entirely manual labour. Also, Prints can fade after several washes.

Cost: Moderate

As you can see, there's more than money and time at stake when choosing the right printing method for your business.

10.5.1   Choose your printing method wisely

Whether you want to scale your business using DTG or screen printing is completely up to you. Before you make your final decision, consider:

- the complexity of your designs,
- your product quantity needs,
- your willingness to invest in the stock upfront.

10.5.2   3D printing in textile industry

Traditional fabric is essentially two-dimensional – strands are arrayed horizontally, vertically, and in crisscross to form a weave. Asfour–who has degrees in mechanical engineering and architecture from the University of Maryland–had a vision, along with Donhauser and Gil, to create "three-dimensional interlocking weaves," which they would achieve with the help of laser cutting. The desire to mess with fabric's third dimension drew them naturally to 3D printing.

The 3D printing belongs to the rapidly emerging technologies which have the chance to revolutionize the way products are created. In the textile industry, several designers have already presented creations of shoes, dresses or other garments which could not be produced with common techniques. 3D printing, however, is still far away from being a usual process in textile and clothing production. The main challenge results from the insufficient mechanical properties, especially the low tensile strength, of pure 3D printed products, prohibiting them from replacing common technologies such as weaving or knitting. Thus, one way to the application of 3D printed forms in garments is combining them with textile fabrics, the latter ensuring the necessary tensile strength.

Several methods enable the additive production of 3D forms, such as stereo-lithography, selective laser sintering or fused deposition modelling (FDM) . The FDM technique is based on resistively heating filaments in an extruder nozzle; the molten material is deposited on the printing bed line by line and then hardens there. The second layer is printed after the first one is finished and the printing bed is lowered .

3D printing in the textile industry lets you unleash your imagination in order to quickly create new structures through innovative new materials.

Thanks to the wide choice of materials that we offer such as plastic or our new flexible plastic material (TPU), it is possible for you to 3D print your most ambitious projects in less time and associate them with clothing of any type to reinterpret the most classic styles.

Marrying different types of fabrics and 3D printing allows you to explore new facets in fashion and therefore to propose a new vision in the textile sector. In an avant-garde sector such as fashion, mixing the latest trends with the latest 3D printing technology will allow you to differentiate yourself from competitors.

10.5.3　Tailor-made with 3D printing

In the textile industry, we are often forced as consumers to choose between different sizes which, depending on the brand, will not be perfectly adjusted to the dimensions of our body.

With 3D printing, you can create custom-made clothing. Various 3D modelingsoftware allow you to create clothing that suits you depending on your body type. You will have to transcribe your measurements on 3D modeling software and thenmodel what you have in mind.

The frames will be tailor-made so that you can print in rigid plastic or flexible plastic. One of these two materials associated with the conventional textile will allow you to create innovative clothing and perfectly suited to your body type.

10.5.4　The flexible plastic: new material to the textile industry

Thanks to our materials engineers we managed to 3D print a new material, the flexible plastic, among the most flexible in the world. This material is thermoplastic polyurethane technology designed specifically for Selective Laser Sintering (SLS) with a level of Shore 65A which makes it highly flexible.

With this new type of material, you are going to reinvent the textile industry by inventing fully 3D printed clothes. The flexibility of flexible plastic allows you how to create clothes with the most unusual shapes while keeping the flexibility of the fabric.

FDM printers can handle filaments produced from several materials, such as Acrylonitrile butadiene styrene (ABS), polyactic acid (PLA), polyamide, polycarbonate, polyethylene, polypropylene, or wax . Additionally, special materials are available, e.g. wood or brick, which are especially useful for model makers in architecture. Soft materials can cause problems in the transport mechanism . One possibility to overcome these problems is using special filaments, such as Lay Tekkks, produced by Kai Parthy from CC-Products (Cologne). The parts printed with this hard filament can be put into warm water for a certain time, which dissolves the hard parts and softens the sample .

Since the FDM technique is used in the most inexpensive 3D printers, this method is the most interesting one for small companies, allowing for testing the possibilities of this technology without the necessity of high investments. While previous experiments always concentrated on the possibilities to create textile-based structures from different polymeric materials by FDM printing and to insert fibrous materials in 3D printed forms , this article examines possibilities to print 3D forms directly on textile fabrics.

11

Yarn printing (Space dyeing)

Space dyeing is a process of dyeing of yarn in which multiple colours are applied along the length of each strand of yarn which may or may not repeat after a fixed interval.It is a technique used to give yarn a unique, multi-coloured effect. While a typical skein (a length of thread or yarn, loosely coiled and knotted) has the same colour throughout, a skein of space dyed yarn has two or more different colours that typically repeat themselves throughout the length of the yarn. Space dyed yarn is sometimes referred to as dip dyed yarn. In space dyeing, different colours are printed along the length of the yarn to give yarn a multicolour effect. Yarn dyed with at least two different colours on one skein is called space dyed. When space dyed yarns are used to form fabric, patterns can emerge depending on the length and variation of each colour block. Space dyed yarn is a term that defines the dyeing process in which multiple colours are applied along a length of yarn (usually a continuous filament strand) at intervals. This process is typically used when different coloured yarns are used in the construction of fabrics (e.g. plaids, checks, iridescent fabrics). Generally, to dye the yarn, you place loose skeins in a large enamelled steel vat and pour the dye and mordant over the various loops, dipping as necessary to produce the varying colours.

Yarn printing is also known as "Space Dyeing". Although the printing of yarns for true patterned effects proved very difficult to control, the random space-dyed effects that can be more readily attained by a variety of yarn-printing methods have continued to be popular.

11.1   Yarn printing (Space dyeing) techniques

Space dyeing can be achieved through several methods:
1. Knit-De-Knit Process
2. Warp Yarn Printing/Dyeing Process
3. Pot Skein Dyeing Process
4. Package Form Dyeing Process
5. Hank Form Dyeing Process
6. Continuous Dyeing Process

11.2   The knit-de-knit process

In the knit-de-knit process, the yarn is first knitted into tubular fabrics (socks) and then overall length of all loops is dyed to a solid colour. All solid-coloured dyed loops of tubular knitted fabric are dyed with up to seven different colours. After dyeing, tubular fabrics are passed through an ager, where they are steamed for 3-10 minutes in saturated steam. Mordant is used to chemically bond the dye with skein. This is followed by rinsing, washing, finishing and drying.

The socks are then de-knitted, producing contrast effect between overprinted and base colours because the overprinted colour does not penetrate at the yarn crossing points.

These yarns usually have short (1/8-1/4 inches) spaces of colour. In this process, the dye penetrates the loops of the yarn but since it does not readily penetrate the areas of the yarn where it crosses itself, alternating dyed and undyed spaces appear. In other words In the knit-de-knit process, the yarn is first knitted into a tubular fabric (sock), dyed to a solid colour and then overprinted with up to seven different colours. The socks are then de knitted , producing contrast effect of overprinted and base colours , because the overprinted colours do no penetrate at the yarn crossing points. These yarns usually have "short" (1/8"–1/4") spaces of colour.

Knit–deknit applications, on the other hand, tend to give characteristically speckled 'microspaced' designs, because of the limited degree of penetration of dye liquor achieved by the duplex printing rollers into the yarn sock.

Although the end-effects produced by the two methods are basically different, the processes can be modified so that their results are more closely comparable. Thus the long spacing effects of warp printing can be imitated by overall application of a ground colour followed by colour spotting with segmented lick rollers or oscillating jets of dye liquor. Similarly oscillating jets of liquor can be applied to knitted sock and excess liquor squeezed out before steaming. This leaves large coloured areas with good liquor penetration and, when tufted, the long spacing effect is achieved.

11.3   Warp yarn printing/dyeing process

Warp yarn printing is roller printing applied to multiple strands of yarns from the warp system which are continuously printed at specific space intervals with different colours. These yarns have usually long spaces of each colour. Typical colour length is 3-7 inches.

Warp printing is roller printing applied to warp yarns before they are woven into a fabric. Fine white or neutral coloured filling yarns are generally used for weaving so that the design on the warp will not be obscured. Warp printing is used for expensive cretonnes and upholstery fabrics. Warp yarns can also be dyed with a base colour, followed by spraying of dye colours at specific intervals. Mordants are used to chemically bond the dye to substrate. These yarns when woven into fabric produces eye-catching designs.

11.4   Pot skein dyeing process

Skein is added to the pot (dye bath) having water before turning on the heat. This helps in the reduction of movement of dye in the dye bath when added, enabling maximal separation between the colours. Water present in the pot is boiled and skein is adjusted in the pot with a spoon. Different coloured dye is added into the pot with each drop in a different spatial area. There is no stirring the contents of pot in whole process as we want as little movement in the dye bath as possible. The skein is left in the pot undisturbed until all of the dye has been absorbed into the skein. When skein is cool enough to touch, it is washed with warm water and mild soap until the rinse water runs clear.

Multi coloured skein is obtained in this process with each strand of yarn having random colours on it at random intervals which produces very distinct designs when transformed into fabric. Dye is fixed on skein with the help of mordant.

11.5   Space dyeing in package form

Various package dyeing machines , with a single package dyeing capacity are available in which we can dye up to 8 colours , each colour is having a different feed tank as well as a injector pump. The colour is injected at a fixed place at a high pressure and collected through a vacuum. The resulting package has different patches having different colours which may or may not be at fixed intervals and, hence space dyed yarns are obtained.

This type of process is used to create beautiful patterns of dyed patches on package, with good sharpness. The dyed time is extremely low. The packages are taken for further colour fixation and washing on another set of machines such as autoclave and yarn dyeing machines.

Various package dyeing machines , with a single package dyeing capacity are available in which we can dye up to 8 colouurs , each colour is having a different feed tank as well as injector pump . The colour is injected at a fixed place at high pressure and collected through vacuum .

This type of machine is used to create beautiful patterns of dyed patches on packages , with good sharpness .The dyed time is extremely low . The packages are taken for further colour fixation and washing in another set of machines such as autoclave and yarn dyeing machines.

11.6   Space dyeing in hank form

This is similar process to produce tie dye (Tie dye is a technique for dying natural fabrics that results in interesting, colourful patterns) effects in knitting yarn and embroidery threads.

The processes which can be employed for space dyeing in hank form are:

By spraying the colours on a layer of hank at fixed places, which are fixed through hank with pressure and collected underneath by vacuum. The yarns are dried, and cured for colour fixation and the washed.

The roller dyeing machines can be used very effectively for space dyeing with vat and naphthol colours. Even reactive dyeing is also possible with a slight modification of machine for mercerized cotton as well as rayon.

Space dyeing machine with fibre carrier can be used for space dyeing in polyester hank, but the method involves a higher labour cost. But desired effects can be obtained using this method. Dye is fixed on the hank with the help of mordants.

This is a simpler process to produce tie dye effects in the knitting yarns and embroidery threads.

11.7   The continuous dye process

In the continuous yarn dyeing process, the yarn is dyed as a single or plied yarn and colour is applied by air jet. This process allows for yarn to have either long or short spaces of colour.

The patent literature abounds with systems for producing coloured flecked effects on yarns but the two most successful methods entail either warp printing or colour application to a tubular knitted 'sock'. The essential process sequence begins with dye liquor application, followed by steam fixation, washing-off and drying .

Various warp-printing methods have been used over the years. In most present-day systems several ends of carpet yarn are taken from wound packages on a creel and colour is applied, either by lick rollers or by some form of spray or spinning disc applicator, to the yarns. The yarns are carried past the print heads in warp form or lying on a brattice on which they have been laid down in a continuous circular or elliptical coil. Warp printing tends to give the so-called 'long spacing' designs in the tufted carpet produced from them.

In contrast, the Crawford Pickering warp-printing system was designed to produce fully patterned tufted carpets with up to eight colours. The warp of yarns was passed between a pair of cylinders, around the surface of which were mounted rows of small dye applicator pads (about 20 mm square). The lower cylinder dipped into a trough of dye liquor and dye was thus picked up on the surface of the pads. These pads could be actuated mechanically so that when opposing pairs were in the raised position the yarn passing between them was printed. With a typical 5 mm pile height carpet, a 20 mm printed length of yarn was equivalent to two tufts of the final carpet. With longer pile carpets of the Saxony or shag pile type single tuft definition was possible, and it was on such carpet constructions that the best results were ultimately obtained. In the original machine the pattern control mechanism was a movable notched bar, the positions of the notches determining the raising or lowering of the print pads. Preparation of pattern bars was therefore rather tedious; later, one company introduced electro pneumatic actuation of the individual print pads with pattern data provided from a microprocessor. Ultimately, however, the full patterning potential of this machine was not realized, mainly because of the problem of keeping a warp of printed yarns in register. There is now only one machine left in operation.

Index

Printed in the United States
by Baker & Taylor Publisher Services

Printed in the United States
by Baker & Taylor Publisher Services